Fell's

Help Your Child Excel in Math

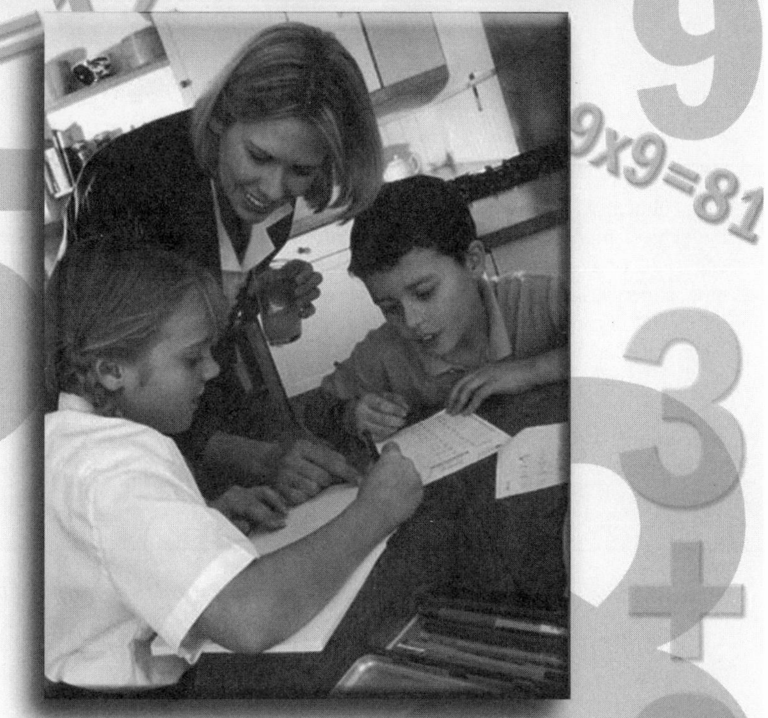

- Enhance Classroom Work
- Play 34 Fun Math Games & Puzzles to Maximize Retention
- Teach skills in a Logical, Developmental Order
- Give Your Child the Fredoom to Learn
- Information on What Questions to Ask at Parent-Teacher Conferences

MARGARET BERGE & PHILIP GIBBONS

First edition. 1992
For the purposes of clarity and consistency, all references to gender appear in the masculine form.

Library of Congress Cataloging-in-Publication Data

Berge, Margaret.
Help your child excel in math / Margaret Berge & Philip Gibbons.-- 2nd ed.
p. cm.
ISBN 0-88391-065-9 (pbk. : alk. paper)
1. Mathematics--Study and teaching (Elementary) I. Gibbons, Philip. II. Title.
QA135.6.B47 2004
649'.68--dc22

2004003180

1 0 9 8 7 6 5 4 3 2
Manufactured in the United States of America

Dedication

This book is dedicated to those parents willing to give their children the freedom to learn.

- The Authors

We wish to extend our appreciation to Marilyn Pearson for her contributions to Chapter One of this book and to Jeanne Bolick for contributions to Chapters Six and Nine.

Margaret Berge
Philip Gibbons,

Drawings by Nycia Countryman

HELP YOUR CHILD EXCEL IN MATH

CONTENTS

Introduction xvii

PART 1

1. PREPARING TO USE NUMBERS 1-19
    - Words Naming Position and Direction 2-3
    - First, Next, Last 3
    - Sorting by Likenesses and Differences 4
    - Comparison and Order by Size and Weight 4-5
    - Shapes: Identifying and Drawing 6-7
    - Patterns 7-9
    - One-to-One Matching 9-10
    - Counting: One to Ten 10-13
    - More and Less 13-15
    - Readiness for Addition and Subtraction 15-16
    - Numerals: 0-10 16-17
    - One Half 18
    - Summary 19

2. USING NUMBERS TO TEN 21-37
    - Counting and Writing Numerals to Ten 22-27
    - Ordinal Numbers: 1st to 10th 27-28
    - Money: Pennies, Nickels. Dimes 28-29

Measuring Length with Inches and Centimeters 29

Addition and Subtraction Facts with Sums to Ten 30-37

Summary 37

3. USING NUMBERS TO 99 39-60

Counting and Writing Numerals to 99 40-46

Fractions: One Half, One Third, One Fourth 46-49

Addition and Subtraction Two-Digit

Numbers without Regrouping 49-50

Time on the Hour 50-51

Addition and Subtraction, on Facts.

Sums Eleven to Eighteen 52-57

Graphing 57-59

Probability 59-60

Summary 60

4. USING NUMBERS TO 999 61-83

Counting and Writing Numerals to 999 61-64

Addition and Subtraction: 3 digits without Regrouping 64-65

Fractions:Halves,Thirds,Fourths 66-68

Multiplication and Division through the Five Facts 68-76

Telling Time 76-77

Liquid Measures 77-79

Addition and Subtraction: Two-digit

Numbers with Regrouping 79-83

Summary 83

5. USING NUMBERS TO 9.999 85-104
Addition and Subtraction: Three-digit
Numbers with Regrouping 86-89
Fractions: Equal to One. Equivalent, Comparing 89-92
Multiplication and Division through the Nines Facts 92-93
Multiplication: Two- and Three-Digit Numbers by
One-digit Numbers 94-96
Division: Two- and Three-digit Numbers by
One-digit Numbers 96-99
Measurement of Height 99
Measurement of Length 100-101
Perimeter 101-102
Area 102
Thermometers 103-104
Summary 104

6. EXTENDING THE USE OF NUMBERS 105-127
Place Value 106-109
Computation 109-118
Adding and Subtracting Fractions with
Like Denominators 118-120
Multiplying Fractions 120
Solving Two-Step Problems .... 121 -122
Finding Perimeters of Polygons 122-123
Finding Areas of Rectangles and Other Shapes 124
Finding Volume 124-125
Miles and Kilometers 125-126
Scale Drawings 126-127
Summary 127

PART 2

7. MONEY: EARNING IT, SPENDING IT. AND SAVING IT 129-142
Earning It 130-133
Spending It 133-140
Saving It 140-141
Summary 141-142

8. MATH AND THE NEWSPAPER 143-150
Advertisements 143-147
Sports Section 147
Homes Section 147-150
Summary 150

9. MATH AND TRAVELING 151-163
Making Short Hops Around the Neighborhood 151-154
Planning Longer Trips 154-157
Making Longer Trips by Car 157-162
Traveling by Plane, Train or Bus 162
Summary 163

10. MATH POTPOURRI 165-176
Sports 165-168
Hobbies 168
Personal Math 168-171
Fun with Math 171-173
Calculators 174-176
Summary 176

11. GAMES 177-199

CONCLUSION 200

Help Your Child Excel in Math

- Enhance Classroom Work
- Play 34 Fun Math Games & Puzzles to Maximize Retention
- Teach skills in a Logical, Developmental Order
- Give Your Child the Fredoom to Learn
- Information on What Questions to Ask at Parent-Teacher Conferences

INTRODUCTION

1, 2, 3, 4, 5, 6, 7,
8, 9 . . .

To PARENTS:

Opportunities abound in the home for the practical use of mathematics. Activities described in this book capitalize on these opportunities to develop and reinforce your preschool and primary school child's math skills.

The approach is three-fold:
1) To guide you as you introduce basic math concepts to your preschool child;
2) to reinforce and enrich the math that your child learns in school by providing opportunities for him to solve problems in daily life outside the class room; and
3) to provide suggestions to enrich the child's homework assignments and motivate him to memorize facts.

Parents who have limited time to help their child will find the activities in this book helpful for creating a mathematically-stimulating environment in the home. Most materials suggested for "hands-on" experiences are common around the home, and there is little need to prepare any specific materials. Ways to have the child practice math during the daily routine are emphasized.

Often, this can be accomplished merely by structuring a conversation and asking questions that stress the math applicable to a situation.

For example, the simple act of setting the dinner table can provide practice for several math skills.
1) The precounting skill of matching objects one-to-one. Say: "Place one fork beside each plate on the table."

2) Counting: "Count the plates to find how many napkins we need. (Pause until the task is completed). Now, count the napkins. (Pause). Place the napkins on the table."
3) Identifying left and right: "Place one fork to the left of each plate. (Pause). Place one knife to the right of each plate."
4) Finding fractional parts: "Fold each napkin into halves, and then fourths."

Part I of this book, "Math in the Primary Years," follows the sequential order of most textbooks. The introduction for each skill in Chapter 1 tells the approximate age when your child might be able to learn the skill, or suggests ways to determine if the child is ready to learn it.

The introductions to Chapters 2 through 5 describe the grade level in which the skills in each respective chapter arc usually taught. Chapter 6 describes patterns and activities to help your child extend the use of numbers beyond that commonly learned in the primary grades. Let the child's teacher determine if he is ready for these skills.

Part 2, "More Math in the Home," emphasizes activities that arc common in children's lives outside the classroom and treats the math skills as incidental to the activity. Read Chapters 8 through 10 and become familiar with the math that you can incorporate into each activity, depending on the child's interests and math background. Chapter 11 contains games that are designed to reinforce math skills presented throughout the book.

It is extremely important for your child to develop a positive attitude toward mathematics. If he has not mastered certain skills, he may have an "I can't do math" attitude. This insecurity is commonly known as math anxiety, and is often inadvertently reinforced at home by a parent who says, "I never was good in math myself." It is essential that you do not transfer these negative feelings to your children. To prevent your child from developing this fear of math, follow the guidelines described in this introduction. Let him know that there are unlimited opportunities for both boys and girls in math and its related fields.

Be alert to all opportunities to involve your child in mathematical problem solving and encourage him to name the process (addition, subtraction, multiplication, division) that he must use to solve problems. This step is necessary even if he uses a calculator to find the answers.

The oft-neglected area of recreational math can be educational as well as entertaining. Though this book presents a variety of these thought-provoking activities, encourage your child to try to solve other puzzles and play other games that require the use of logic, reasoning and basic math skills. Many board games (checkers. Monopoly) and card games involve these skills. Written activities which ask for needless repetition, such as, "Work these 50 multiplication problems." often do more harm than good. The child may gain some speed and accuracy from this practice, but, consequently, may develop a strong dislike for mathematics. So let the teacher determine when these activities are necessary.

Your child will undoubtedly be affected by developments of the computer age. The availability of hand-held calculators will have a significant influence on how he solves arithmetic problems. Although these machines provide great speed and accuracy, they do not replace the child's need to understand arithmetical processes. The calculator, as other current computers, can only do what it is told; therefore, the child must be able to tell it what process to use to solve any problem. A special section of Chapter 10 is devoted to math in the computer age.

Jean Piaget, a noted child psychologist and educator, emphasized that a child must have the opportunity to do things over and over again, thus assuring that what he has discovered is true. He also emphasized that practice should be enjoyable and result in success, so follow these guidelines as you use the activities in this book:

1) Read the topics at the beginning of each chapter in Part I, compare with your child's homework, or consult the teacher to determine the skills that the child should be able to apply to situations at home. Then choose the appropriate activities from the book. Do not push ahead of the child's place in the math curriculum, or both of you may become frustrated.
2) The suggested activities are guides only. You may discover similar activities that will be better adapted to your home environment and your child's interests.
3) Ask your child many questions and listen to the responses. Do not label the answers as wrong or silly, but encourage the child to explain them.
4) Encourage the child to ask questions and to comment on his experiences.
5) Let the child see the ways that you use math daily (measuring ingredients for cooking, counting change, following a budget, telling time, etc.).
6) Note that the material in this book covers the math that your child will encounter during the preschool years (ages four or five) through the third grade; this covers a period of four to five years. It is therefore essential that you follow guideline #1 above.
7) Study your child's homework assignments and how he completes them. Use the activities in this book to supplement, not replace, those methods introduced by the classroom teacher.
8) Be patient. Praise effort as well as success. Never scold the child for not knowing how to complete a task.
9) Keep work periods short. Length of attention spans vary from child to child, age to age, and according to the child's interest in the activity, so be alert for your child's waning interest and stop when either you or the child becomes tired.
10) Pick and choose activities. If several are suggested for the same skill, repeat those that the child enjoys most. Skip those in which he shows no interest.

11) Skip activities that involve physical skill that the child may not possess, such as bouncing a ball.
12) Use activities in which the child performs better than you. such as jumping rope. He will love helping you.
13) Continue to use activities designed for one skill until the skill is mastered. Several short periods of involvement are more effective than one long session, so keep coming back to it. Parents tend to go too fast
14) Try to select times when your child is rested and in a positive mood to involve him in these activities.
15 Don't be too quick to tell the answers. Let the child experiment and find them for him or herself.

As children learn to walk and talk at different ages, so will they learn mathematical processes at different ages. Your patience and acceptance of the child as an individual will help determine the degree of personal success in mathematics at home and at school.

PART 1

PREPARING TO USE NUMBERS 1

Children learn faster, say most psychologists, before the age of six than at any other time of their lives. These preschool years, therefore, are (he ideal time for you to begin introducing basic math concepts to your child. Take this opportunity to guide his learning and to encourage a positive attitude toward math by providing interesting activities at home.

Your preschool child is learning by doing-by trying things out to see what works and what does not. This active involvement helps the child understand, as well as remember, what has been learned, so direct this curiosity and energy to help the child understand and remember math concepts.

Parents who have had unpleasant experiences in math can transfer negative feelings to their children, without realizing the effect, by making comments about their own disdain for the subject. This can create a feeling often called math anxiety. Avoid this pitfall by approaching all activities with enthusiasm and objectivity. Follow the guidelines described in this book to provide a stimulating, supportive learning environment in your home. Your involvement and interest in your child's math, as in any part of his life, convey that you think these activities are important and deserve the child's greatest efforts.

This chapter contains many activities to foster your child's basic math skills during the preschool and kindergarten years. Children learn at different rates, so we have suggested an age range at which many begin to learn each concept. Allow the degree of your child's interest in the activities to determine when he is ready to learn the related skill.

This chapter includes the following:
1) Vocabulary builders:
2) math concept builders;
3) an introduction to number and numeral recognition: and
4) an introduction to money.

The child does not have to be able to write to do any of these activities. Save the writing experience for the kindergarten or next grade teacher to introduce.

Words Naming Position and Direction

As soon as your child begins to ask questions about, or discuss, the size. shape or position of figures in space, he is using the branch of mathematics known as geometry. Learning the spatial relationships and matching vocabulary that is presented in this chapter will prepare the child to follow directions, written and oral, and to reason logically when completing a task or forming an idea. Most children learn to use the following words by the age of five. Use these words often in conversations with your child.

- Words thai name position: above, below; inside, outside; over, under. beside, on. behind; near. far.
- Words that name direction: left, right; up, down; forward, backward; higher. lower.

Ask questions or give simple directions that contain the words and see if the child follows them. Limit the number of new words in each conversation to one or two and be ready with help and reminders if he seems confused. Add new words only when the child is ready for them.

PLAY: Simon Says. Game 1; 1 Spy, Game 1; Hide the Thimble; Bean Bag Toss (all described in Chapter 11).

POSITIONAL WORDS: (above, below, over. under, beside, on. behind, near, far. nearer, farther.) Ask your child to name something that is far away, and then something that is near. If he doesn't know, stand close and say, "I am near you." Move to the opposite side of the room and say, "Now I am far from you." To practice using the words nearer and farther, play Bean Bag Toss (from Chapter 11).

If your child cannot readily tell you what object is beside, above, below, on or behind another object, show and tell him. Then ask questions that will encourage the child to use the same words. "The apple is on the table and the kitten is below the table. Where is the cup? Yes, the cup is on the table. Name something else that is below the table. Yes. the kitten's dish is below the table."

DIRECTIONAL WORDS: (forward, backward, up, down, left, right, higher, lower.) Include directional words in activities with your child. Your directions will involve movement and the child probably has plenty of energy and is ready to be up and about. To reinforce the directions and to show the child that the words in each pair are opposites, use activities similar to the following:

Tell your child: 'Take a little jump forward. Take a big jump backward. Run forward. Run backward. Put your hand up. Put your hand down. Sit down. Stand up."

Two other directional words that are important to your child are left and right. Use these terms often in directions and questions. Tie a ribbon around his right wrist and explain that the right wrist has the ribbon and the left wrist has no ribbon. Then give directions, such as: "Put on your right shoe. Wash your left arm. Try to rub your nose with your left elbow." When you are setting the table, ask the child to put a napkin to the left of each plate; a knife to the right of each plate; a spoon to the right of the knife. After he understands each word individually. put both words in one activity: "Scratch your left ear with your right hand."

First, Next and Last

Taking turns is probably a daily occurrence around your house. Take advantage of this usual happening to introduce these terms of order-first, next and last-to your four or five year old.

Examples:

1) When the children play games, have them identify the players as first, next and last in turn.
2) When you serve snacks, ask a child to tell who is to get the first, next and last servings.
3) When discussing storybook happenings, ask questions about the sequence of events: "Whose chair did Goldilocks sit in first? Next? Last?"
4) Ask questions similar to the following about the child's day: "What do you do first when you get out of bed in the morning? What do you do next? What do you do last before you go to bed at night? Which do you put on o first, your shoe or your sock? What is the last food you ate at dinner?"
5) Use the words when giving directions. "Put the plates on the table first Next, put the napkins on the table. Place the silverware last."

Sorting by Likenesses and Differences

Each of us performs classification tasks daily. We separate tableware into sets according to likenesses and differences; we classify animals as cats, dogs, cows, horses; we choose an outfit to wear by matching colors; we locate a house by classifying the address as an odd or even number. You can bring order to your four or five year old's environment and strengthen the child's math concepts of size, shape and number by providing opportunities for the child to classify objects by likenesses and differences.

You probably have many materials that arc appropriate for these activities. Adjust the following activities to the material and give as much guidance as the child needs to successfully complete the task.

1) Show a set of items that are different in only one way-i.e., size, color or shape. Ask the child to put the items together that are alike. Example: large and small paper clips. If the child has difficulty, place three large paper clips in a set and three small paper clips in another set and say: "Put the other paper clips in the set that they match. Other examples: buttons of two colors that are alike, except for color; knives, forks, spoons; dinner forks and salad forks; serving spoons, soup spoons, teaspoons; nails of two sizes; two kinds of tools in the workshop.
2) Display a set with items that are different in two ways: large red buttons, and small red buttons, large blue buttons and small blue buttons. Ask the child to put the ones that are alike together and notice how he groups the items. Ask how the items in each set are alike, and how the two sets are different.

Comparison and Order by Size and Weight

When children four or five years old begin to compare objects by size and weight, use only two measurements at a time to prevent confusion. Introduce the vocabulary gradually, and provide varied experiences for each set of terms. Some of these terms are: small, smaller, large, larger, short, shorter, long, longer. tall, taller; heavy, heavier, light, lighter. Begin by using pairs of objects that are greatly different in length or weight so that the child can easily distinguish the difference. Give directions and ask questions that involve the terms.

Examples:

1) Show a large and a small ball (beach ball and tennis ball) and say: "Let's play with the smaller ball." Let the child select the ball.
2) Show two pieces of fruit. Ask the child "Do you want the larger or smaller piece?" Notice whether or not he takes the piece indicated.

3) Show two books of greatly different weights and say: "Lift each book and hand me the heavier one."
4) Show two packages and say: "Let's wrap the heavier package with the blue ribbon and the lighter one with the white ribbon. Put the blue ribbon with the heavier package. Where will you put the white ribbon?" (With the lighter package.)
5) Ask which of two pieces of string is longer. Then say: "Please hand me the shorter string."
6) Have two children stand side by side in front of a mirror. Ask who is taller, or shorter.

After the child is able to compare the size of two objects, have him use the following words to describe the comparisons of three or more objects: smallest, largest, shortest, longest, tallest; lightest, heaviest

Here are activities for ordering by length or height:

1) Have three people stand in a staggered fashion in front of a mirror so that they look like stairsteps. Ask the child to name the tallest and shortest per sons. Discuss the fact that they are in order from shortest to tallest, left to right or right to left
2) Ask the child to arrange plants on the window sill from shortest to tallest, left to right. Mention that they look like stairsteps.
3) Give the child three or more paper dolls, dolls, or stuffed animals of varying heights and three outfits (one to fit each). Ask the child to arrange the dolls or animals in order, from shortest to tallest, and to give each the outfit that fits.
4) Using the dolls or animals and three other outfits out of order, indicate one doll and ask the child to choose an outfit that fits it. Repeat this activity until the child realizes that the selection will be easier if he arranges the dolls and the outfits in order, according to length.
5) Ask your child to arrange some ribbons (string, shoelaces, paper towel tubes, straws, etc.) of different lengths in order by length. Start with items that are very obviously different and move to items thai are closer to the same length.

Here are activities for ordering by weight:

1) Ask your child to arrange three or four packages in order from lightest to heaviest. Ask if he can tell the lightest by looking. (No.) Ask how he can find it. (Lift each package.)
2) Prepare three small boxes (jewelry boxes, paper clip boxes, etc.) of the same size by putting rocks in one, paper clips in another and nothing in the third. Have the child arrange them in order from lightest to heaviest or heaviest to lightest from left to righT (Play I Spy, Game 2, from Chapter 11.)

Shapes

Your child is probably developing a keen interest in his surroundings and beginning to son things out, to see likenesses and differences in the shapes around him One step in bringing order to the environment of your four or five year old is to help him identify four of the most common shapes by name: circle, triangle, square and rectangle.

IDENTIFYING SHAPES: Help your child hunt for objects that can be traced to make circles, rectangles, squares and triangles. Let the fun of the hunt be the major objective of this activity. Search for one shape at a time so that each is distinct in the child's mind.

Circles will be easy to find. Some suggestions are: drinking glasses, jars, jar lids, saucers, rings (anything that is perfectly round, not oval or egg-shaped). Rectangles and squares will not be too difficult to find. Spice cans. books, boxes, plaques and pictures, tiles, toy pieces and blocks will all serve. Mention that a square is a special kind of rectangle, that all four sides of the square are all the same length. Place two squares of the same size one on top of the other. Turn the top one a quarter-turn. Point out that they still fit together and. therefore, all sides are the same length.

Finding triangles will require some luck. Perfume bottles, scraps of wood or cloth into pieces for a puzzle may provide these shapes. If your search produces nothing, make a traceable triangle shape by cutting off a comer of a cardboard box or tracing two sides of one of your rectangles and drawing in the third side. Try to have differently-shaped triangles so that the child can see that every three-sided shape is called a triangle.

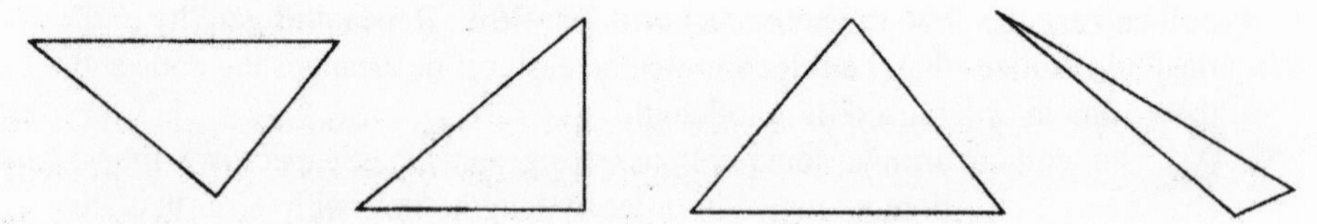

After you have collected objects in all shapes, talk about each shape and its attributes, i.e. a circle is round, a triangle has three sides, a rectangle has four sides, and a square has four sides of the same length.

Trace the shapes, color them. cut them out, and hang the prettiest one of each shape. Make a mobile by hanging the different shapes from a coat hanger, using lightweight string or fishing line.

Continue to encourage your child to look for the different shapes wherever he goes. Look for them in road signs, at the grocery store, in magazines, at friends' homes, in architectural design, etc.

DRAWING SHAPES: Approach this activity with an enthusiastic, "Let's draw! I feel like drawing circles, triangles, rectangles and squares. Let's draw pictures using only these shapes."

Suggestions:

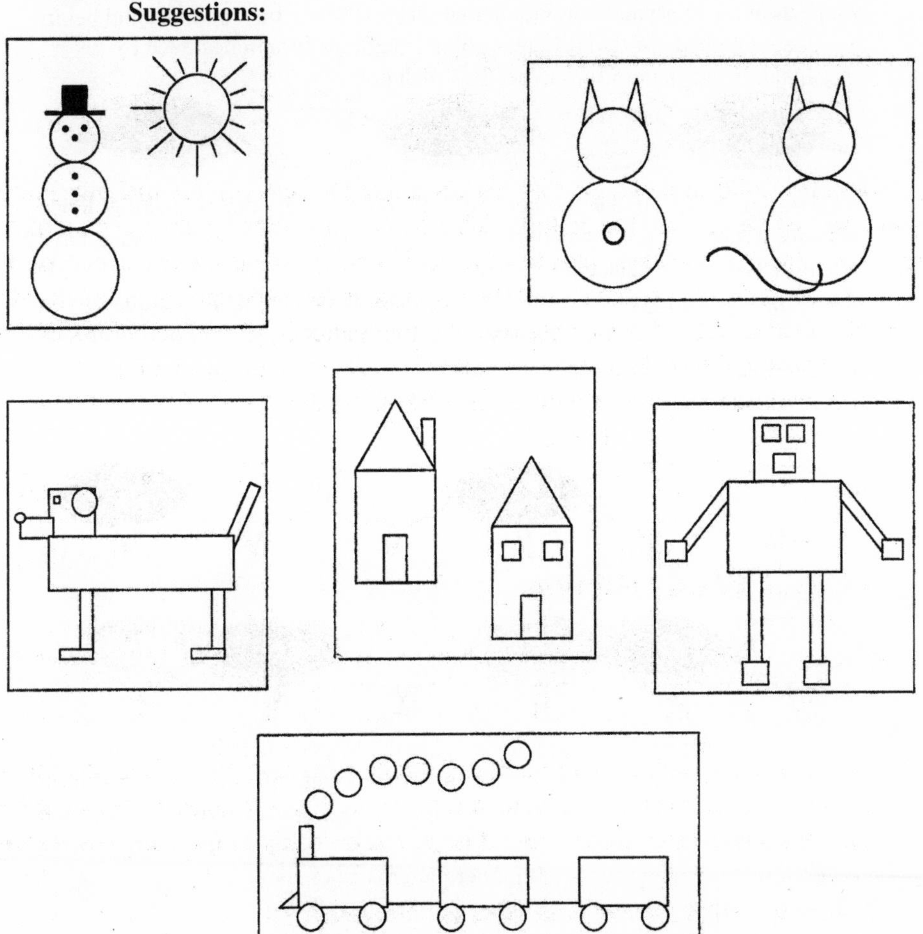

After you have drawn the pictures, ask the child to point to a circle, to a square, to a rectangle and to a triangle. Play Shapes from "Games" in Chapter 11.

Patterns

Number relationships and patterns in our number system are fascinating. They also provide shortcuts for computation, methods for checking accuracy, and just plain fun.

Increase your child's readiness for discovering these patterns and number relationships by helping him become aware of patterns in the environment. The following activities are examples of ways to guide your child as he develops skill in repeating a pattern and continuing one that is already begun. Examples:

1) Use snap beads of two different colors. Begin to form a necklace by repealing the pattern of one bead of each color.

Ask the child to make another necklace like the one you are making. Place yours so that he can see the pattern. After he has put some beads together, have the child place the necklace beside yours and check to see if the two are alike. If they are, let the child continue until the necklace is long enough for either of you to wear. If there is a difference between the two patterns, place each of his beads close to yours and help find the bead that will make the patterns match.

Follow the same process with other patterns, such as:

2 white beads and 1 red bead

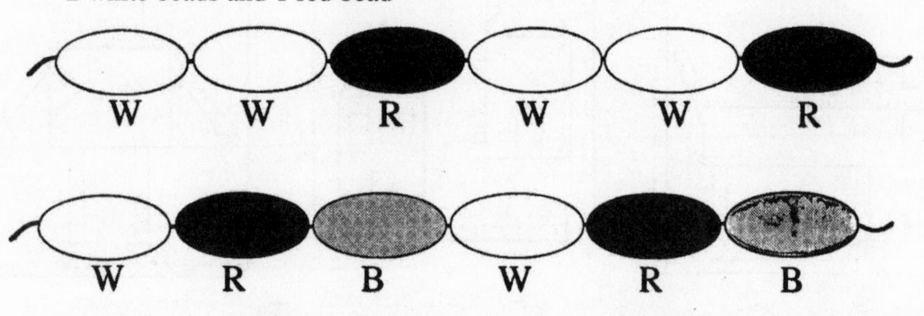

2) Begin another necklace with a new pattern. After completing the pattern at least twice, ask the child to hand you the bead that will come next. See if he can pick the correct color without help. but be ready to give any assistance he needs.

Follow the same process with other patterns, such as:

3 red, 1 white

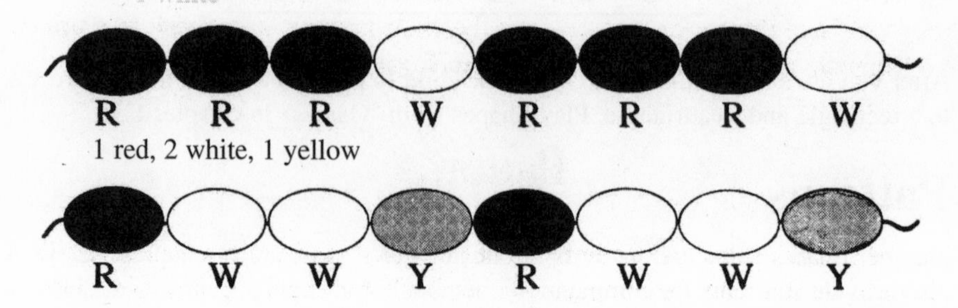

1 red, 2 white, 1 yellow

3) Use blocks from the toy box to form patterns. Begin the pattern and let the child complete iL
4) Use shapes cut from paper or cardboard to form patterns and let shape be the determining characteristic, instead of color. Ask the child to complete (he pattern.

Look for other items around the house that can be used for forming patterns. Checkers and buttons are good possibilities. Also, use size as a characteristic. Example: make a chain with safety pins or paper clips.

1 large, 1 small...and so on.

Reverse the roles, and allow the child to make the pattern for you to complete. Your child might enjoy looking through necklaces for patterns and reproducing them.

One-to-One Matching

Before four or five year old children begin to count, they should have experience in matching objects in two sets on a one-to-one basis. The home offers many excellent opportunities for the development of (his skill.

Examples:

1) Setting the table. Place two or three plates on the table and ask your child to:
a) get a napkin for each plate. Many children will get as many napkins as they can carry. Others will take one napkin at a time and place it beside a plate until there is a napkin for each plate. After several of these experiences, the children will make efforts to bring just as many napkins as there are plates. If your child gets too many, or too few. napkins, encourage him to make corrections by returning the excess to the prop* er place, or by getting the extra ones needed.

After he does this, thank the child for getting one napkin for each plate. When the child is able to match the napkins to the plates, one-to-one. use the same activity with more than three plates. Increase the number of plates gradually to keep from confusing the child.

b) get a knife for each plate.
c) get a spoon for each plate.
d) get a fork for each plate.
2) Sharing. When friends are present (two to five children), ask your child to gel a small toy (marble, jack, toy car) for each friend. Watch to see if he has some left over or if she makes several trips to the toy box.
3) Making the beds. Ask the child to bring a pillowcase for each pillow on the bed.
4) Gardening. Ask your child to bring one plant for each hole or container.
5) Arranging the workshop. Show three or four tools and ask the child to bring one hook, or nail, for hanging each tool.

Counting

Many children begin to learn to count at four or five years of age, depending on the amount of guidance provided, but often they are merely parroting the words - one, two, three, four, five. Your child may not learn to count to ten for many months after you begin to use the activities. Be patient, supportive, and enthusiastic.

Before a child can fully understand counting, he must have numerous experiences counting objects that he can feel and move about while saying the matching number names. Again, he is matching one-to-one- one number name to one object. There is too little time available for this handling and moving about in the classroom but the opportunities at home are great. The following pages will provide specific suggestions for helping your child learn to count one through ten. Try these activities after your child is able to match objects one-to-one.

Start with the word one to match one object. Then present each number as one more than the previous number. Follow these procedures to develop the concept of each number.

1) As your child watches, count the objects in a set and tell how many.
2) Have your child count and answer the question "how many" about another set of the same number but with different objects.
3) Have your child produce a set of that same number.

Does this sound time consuming? As your read the following suggestions, you will see how you can integrate the process into your daily routine.

ONE: Here are examples of ways to help your child learn to count one object:

1) Woodworking. Show one hammer and say: "This is one hammer." Show one nail and ask: "How many nails do I have?" If the answer is "One," ask him to bring you one screwdriver (or other tool). If the child has trouble identifying other sets of one, continue to show sets with one member in each. "Thank you, but this is one screwdriver. (Show one.) Please put these

screwdrivers back on the shelf. (Give him the extras.) Now, how many screwdrivers do I have?" (One.)

2) Potting plants. Show one plant and say: "Here is one plant. Please bring one pot for the plant." If he brings one pot, thank the child and ask: "How many pots did you bring?" (One.) If he brings too many, show the child one pot and have him put the extras back on the shelf. Say: "Please bring one bottom for the pot." Respond as before.
3) Sewing. Bring one button to me.
4) Decorating. Put one rose in each vase.
5) Fishing. How many minnows are on the hook?
6) Playing. Bounce the ball one time. Toss the ball and pick up one jack.
7) Dressing. How many belts do you have?
8) Looking at books. How many dogs are in the picture?
9) Saving. Put one penny in your piggy bank.

Continue these types of activities with other objects, asking your child to bring one item and responding to his efforts in the same pattern as described above. Remember! Use every opportunity to:

1) Tell how many are in a set of one;
2) have your child tell how many are in a set of one;
3) have your child pick up a set of one; and
4) play Simon Says, Game 2 from "Games" (Chapter 11).

Two: Stress: One and one more make a set of two. When playing jacks, say: "Pick up one jack when you bounce the ball. Now, pick up one more jack. One and one more make two. So you have two jacks. We can count your jacks. (Move a jack as you say each number.) One, two."

Provide many opportunities for the child to count two objects. Ask for two forks, buttons, pillowcases, tools. Encourage your child to touch each item and count-one, two-then answer the question "how many?"

At play time and book time, use the following activities:

1) Make a game of finding sets of two and making rhymes for counting:

One, two	***One, two***
Your eyes	***Small ears***
are blue.	***on you.***

2) Look at pictures in a book, find sets of two, and ask: "How many trees arc in the picture? How many kittens?"
3) Bounce a ball two times as the child counts.
4) Encourage your child to toss the ball and pick up two jacks.
5) Play Simon Says, Game 2 and Mystery Box, Game l(Chapter 11).

THREE: When you are baking or cooking, say: "Put two apples on the plate. Count them. Now put one more apple on the plate. We can count to find how many." Move one apple as you say each number. "One, two, three. Now you move the apples back as we count together, one at a time. One, two, three."

Have the child count objects in sets of three as you sew, clean, garden, and perform other activities. Have him touch and move the objects as he counts. At play time or book time, read or tell stories and poems such as "The Three Little Pigs," "The Three Bears," and "Three Little Kittens."

Have the child:

1) Find sets of three and make rhymes;

   One, two, three
   Buttons on me.

   One, two, three
   Flowers I see.

2) bounce a ball three times;
3) toss the ball, try to pick up three jacks and catch the ball after one bounce; and
4) play Simon Says, Game 2 and Mystery Box, Game 1, (Chapter 11).

FOUR: To practice counting four objects, follow the same procedures as for one, two, and three. As the numbers increase, the child will need more practice. Be patient.

Show three buttons. Say: "How many buttons do 1 have? Let's count (Move each button as you count.) One, two, three. Bring one more button. Now let's count them. One, two, three, four. Three and one more make four."

Use activities similar to the following at play time and book time:

1) Find sets of four objects around the house and make rhymes about them.

   One, two, three, four
   My dog's legs
   all touch the floor.

   One, two, three, four
   Pretty shells
   upon the shore.

   One, two. three, four
   Knocks are heard
   upon the door.

2) Use blue and red buttons (poker chips of two colors or black and red checkers) to make patterns of four. Talk about the sets:
   Do not forget to use the activities that you used in counting to one, two, and three.

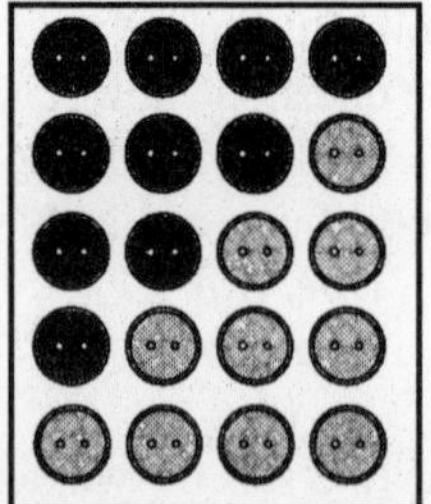

Four red chip*.

Three reds and one blue make four.

Two reds and two blues nuke four.

One red and Three blues nuke four.

Four blue chips

Bounce a ball. Pick up jacks. Play Simon Says, Game 2 and Mystery Box, Game 1 (Chapter 11).

FIVE. SIX. SEVEN. EIGHT. NINE. TEN: Adjust the activities suggested for counting one to four.

Examples:

1) Count the objects in a set of four. Put one more in the set, and stress that five is one more than four. Continue stressing each number as one more than the previous number.
2) After you and the child have developed a number, look for sets of objects and pictures of sets with the appropriate number in each.
3) As the child identifies sets with the appropriate number of objects, make rhymes to match the sets.
4) Use two colors of buttons, chips, or checkers to make patterns of the number.
5) Play Jacks. Take turns with your child. Pick up as many jacks as the latest number developed.
6) Take turns bouncing the ball six to ten times and counting.
7) Play Simon Says. Game 2 and Mystery Box, Game 1 (Chapter 11).

If your child has difficulty bouncing the ball or playing Jacks, praise his effort. If you have trouble with these activities, let the child see you laugh at your own mistakes and be willing to exert extra effort to learn a new skill.

Other activities for developing each number six to ten:

1) Build a staircase with blocks.
Place one more block in each step.
Count the blocks in each step.

2) Let the child jump rope six to ten times and count. This is a great one for you to try. Maybe the child will be better than you. He will love that.
3) Repeat old rhymes, such as:
One, two, buckle my shoe. Three, four, shut the door. Five, six, pick up sticks. Seven, eight, lay them straight. Nine, ten, a big fat hen.
4) Play Checker Cover Up and Clean the Board from "Games" (Chapter 11).

More and Less

Two questions that your child will be asked often in the study of mathematics are: Which is more? Which is less? You can prepare the child to answer these questions. Use the words more and less to talk about two sets of objects that are greatly different in number.

Examples:

1) Show a bag with many beans and a bag with only a few beans. Talk about which bag has more, or which bag has less.
2) Show a bag with a dozen oranges and a bag with one or two. Have the child identify the bag that has more, or less.

3) Show a large box full of crayons and a small box of crayons. Ask the child which box he would like 10 have. Why? (It has more.) Ask which she would rather give away. Why? (It has less.)

After the child has learned to use these terms correctly with amounts that are greatly different, use the terms to talk about sets that are different by just a few. Show two sets of objects and have them matched one-to-one.

Examples:

1) Put three cups and four saucers on the table. Say: "Place each cup on a saucer. Is there a cup for each saucer? (Yes.) Is there a saucer for each cup? (No.) Are there more cups or saucers? (Saucers.) Yes, there are more saucers because there are saucers left over. How many cups are there? (Three.) How many saucers? (Four.) Which is more, four or three? Yes, four is more than three. How many more?" If the child has difficulty telling how many more, ask how many extra saucers there are. (One.)
2) Place five plates and six napkins on the table. Say: "How many plates are here? How many napkins? Is there a napkin for every plate? Is there a plate for every napkin? Are there less plates or napkins? Which is less. five or six? Yes, five is less than six."

Use sets of other objects and similar questioning to continue to reinforce the understanding of the terms more and less.

After your child learns to tell which is more and which is less, even when the difference is very little, ask: "How many more? Less?" Help him match one-on-one to find how many are left over (extra). Use the following activities to see if your child needs practice with one-to-one matching.

1) Place six spoons and four bowls on a table. Ask if there is a spoon for every bowl and let the child check by putting a spoon in each bowl. Remove the spoons that are left over Place the four spoons close together and leave the bowls far apart.

Ask if there is still a spoon for every bowl. If he says no, he is judging on the fact that the spoons occupy less space and he needs more practice matching. If he says yes, ask why. Continue this type of activity and questioning until his responses show that he understands that the same number remains regardless of the space occupied.

2) Show saucers and cups arranged as follows:

Ask if there are more cups or saucers. Have the child check his answer by matching one-to-one.

Develop the terms most and least as you did the terms more or less except that you must use more than two sets of objects when you ask: Which is most? Least? Play Most, More, Less and Least from "Games" (Chapter 11).

Readiness for Addition and Subtraction

A child will learn addition and subtraction facts more easily if he has had many experiences counting real objects in varied situations. These opportunities abound in the home. As you go about your regular duties, supply objects to be counted and ask questions to encourage your child to find the sums and differences. At this stage of getting ready for addition and subtraction, the child needs to see and feel what is happening.

Examples:

1) Have two apples in a basket. Say: "There are two apples in the basket. If you put in one banana, how many pieces of fruit will there be? Try it. Two and one made how many?" (Three.) If two kinds of fruit are joined (apples and bananas) the child will be able to distinguish the two sets and more readily see that two and one make three.
2) As you build in the workshop, say: "I have driven two nails in the board. If I drive one more nail, how many nails will be in the board?" Drive the nails.
3) As you organize the child's room. say: "There is one toy (book) on the shelf. If I put two more toys on the shelf, how many will there be? Try it. Count and tell me how many there are altogether. One and two make how many?"
4) As you garden, say: "I have two tomato plants in the row. If I put three more in the row, how many plants will there be?" Use other activities similar to these.

Since subtraction is the undoing of addition, you can use the same situations to demonstrate this process.

Examples:

1) As you show the fruit basket, say: "There are three pieces of fruit in the basket If you take one out. how many will be left? Try it."

2) Discuss the toys on the shelf: "Three toys are on the shelf. If you take one toy off, how many will be left? Try it."
3) Discuss the tomato plants: "There are five tomato plants in the row. If two die, how many will be left?"

Repeat these activities with two colors of buttons, crayons, pencils, or other items commonly found around the home. Notice that there is no writing at this stage of readiness. Later (probably in first grade) the child will learn to use the signs + and - to write number sentences (equations). These provide a short way to record the stories about what happened when sets were Joined (addition) and separated (subtraction).

Play Mystery Box, Games 2 and 3, and Guess the Number from "Games" (Chapter 11).

Numerals: 0-10

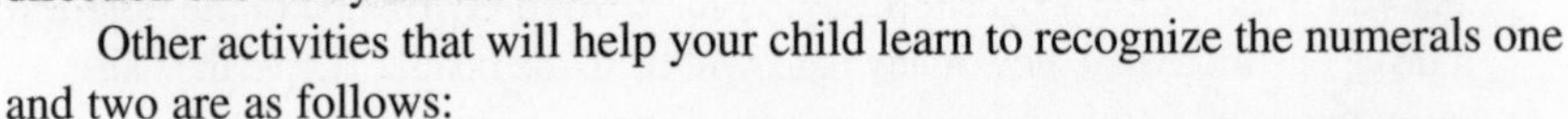

The following activities should help your four or five year old recognize the numerals 1 to 10. Tracing will reinforce his recognition: however, let his kindergarten or first grade teacher leach him to write the numerals.

With a crayon, write the numeral I on a piece of paper. Name the numeral: "One." Encourage the child to trace it with his finger. Be sure that he begins beside the X and moves his finger downward.

Follow the same process with the numeral 2. Encourage the child to trace it, beginning at the point beside the X and moving in the direction shown by the arrow.

Other activities that will help your child learn to recognize the numerals one and two are as follows:

1) Use beans, buttons, or similar objects to make sets of one and two and have the child slip the numeral card that identifies the number under each set.
2) Place each of the cards on the table, face up, next to a pile of beans or other small objects. Have the child make a set on top of the card to match the numeral.
3) Look for the numerals 1 and 2 on signs, in books, on cereal boxes and other containers. Have the child name each numeral.

Your child learns any concept or skill more easily by using as many of his senses as possible. The following activities will use your child's senses of touch and sight to help get the "feel" of the numerals.

1) Cut the numerals from sandpaper and paste each on an index card. Guide the child as he traces the numerals with his finger. The feel of the rough texture will strengthen recognition of the shapes.
2) Let your child form the numerals with a finger in a cookie sheet of salt or sand. Again, the rough texture will provide reinforcement.

After your child has learned to identify the numerals 1 and 2 by name and the number in a set, introduce the other numerals, 3 to 10 and 0, one at a time in a similar manner. As you introduce the new numerals, continue to provide practice of those that he has previously learned. When your child is tracing a numeral, be sure that he begins at the point beside the X and moves in the direction indicated by the arrow.

3 4 5 6 7 8 9 10

The following activities will help your child practice identifying the numerals 0 to10.

1) With a crayon or magic marker, write a numeral in the bottom of each cup of an egg canon. Give the child a pile of dried beans and have him put the number of beans in each cup to match the numeral.

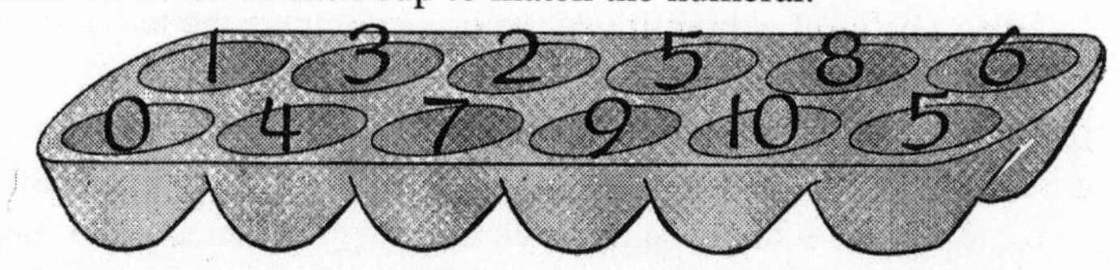

2) Provide a set of numeral cards, 0 to 10, and a set of small objects (beans, but tons, paper stars, etc.) and assist the child as he glues a set of the items on each card to match the numeral. Hang the completed cards in the child's room. Remember, the card with zero will have no objects on it.
3) Now is a good time to leach your child to use the telephone. Write the correct telephone number on the back of a picture of each person that you would want your child to call in case of an emergency. Let him practice calling the numbers under your guidance. This activity will reinforce the habit of read ing from left to right as well as help the child identify the numerals. He will probably enjoy being able to call grandma or Uncle Joe by himself. Do not forget a picture of a policeman and a fireman. If you choose to provide a pic ture of a rescue truck with the emergency number 911. be sure the child real izes he must dial the number only in case of an emergency.
4) Show your child page numbers 1 to 10 in a book. Let her practice finding pages as you give her the numbers.
5) Play Simon Says, Game 3, Match Up, Game 1, and the Numeral Race from "Games" (Chapter 11).

One Half

Before children begin the formal study of fractions, they have many experiences with sharing. Their deep sense of fairness seems to make the concept of halves - two parts of the same size - easily grasped. Use the term one half with these experiences with your four to six year old to stress that it means one of two equal parts.

When sharing an apple with your child, say: "I will give you one half of my apple. I will cut it so that both parts will be the same size. Shall I cut it here? (Show a place that will result in one piece being much larger than the other.) No, I will cut it here." Cut the apple so that the two parts are equal in size. Place one piece beside the other to show the comparison, and say: "The two pieces are the same size, so each is one half of the whole apple. I will give you one half and I will get one half."

Follow the same procedure when sharing an orange or other pieces of fruit. Always stress that if there are two parts of the same size, each part is one half of the whole.

Provide items for the child to share with another person. Ask him to give one half to the other person. Let him divide the item, compare the two parts, and tell you if each part is one half.

To find one half of a set, place 6 peanuts (or any even number of small items) on a plate. Say: "I want to give you one half of the peanuts. Your share must be the same size as mine. I will take a nut each time you take one." After the peanuts are divided, ask: "How many do you have? (Three.) How many do I have? (Three.) Do you have as many as I? Do you have one half? Do I have one half? Yes. each part is one half."

Continue with other sets of two. four. six, eight, or ten objects. Do not expect your child to tell you how many are in one half of a number until he has had many experiences such as (he one above. Continue to use the term one half when referring to the division of a set into two equal parts, but let the child match the objects one-to-one to determine if the parts are the same size. At this stage, the concept of one half as one of two equal parts is the objective rather than naming the number in one half of a set. This will come much later.

Give the child an even number (ten or less) of marbles, pennies or any item that he particularly likes. Tell him to put some of them in each of two boxes and that you will choose the box that you will take. If one box has more than the other, take that one. He will probably begin to divide the sets evenly very soon.

Coins

Who has a better chance than you to give your child many opportunities to handle money? After he learns to count, it is important for your child to have money of his own instead of just getting a handout when he asks for something.

USING A BANK: Your child should have a bank of some kind in which to deposit coins. It will enable you to let money play a greater role in the child's daily life, since real opportunities to spend amounts of 10C will have no value until the child fully understands numbers beyond ten.
With a bank. you and your child can use the following activities:
1) Count pennies as they are deposited;
2) take all of the pennies out and count them in sets often;
3) make stacks of five pennies to trade for nickels;
4) make stacks of ten pennies to trade for dimes; and
5) make stacks of two nickels to trade for dimes.

SHOPPING: Help your child decide how much money he will need before you leave home-stay within this amount. Allow him to gel small items from a machine. If necessary, he can exchange his coins with you or a clerk for the coin needed.

The mechanical horse or other rides outside many stores is always appealing and sometimes costs 25e. Discuss the fact that the ride costs two dimes and one nickel and let the child see if he has these coins. You might show the quarter and explain its value as the same as two dimes and a nickel. But, remember. you've only introduced numbers to 10.

Summary

Chapter 1 has presented activities to help your child master basic math concepts which, if learned well, will enable the child to begin grade one with increased skill and confidence. Teachers are quick to note those students who have received help and encouragement from parents.

Don't assume that these concepts, once learned, will remain intact; children forget. Return to them often and vary the activities as you deem appropriate.

2
USING NUMBERS TO TEN

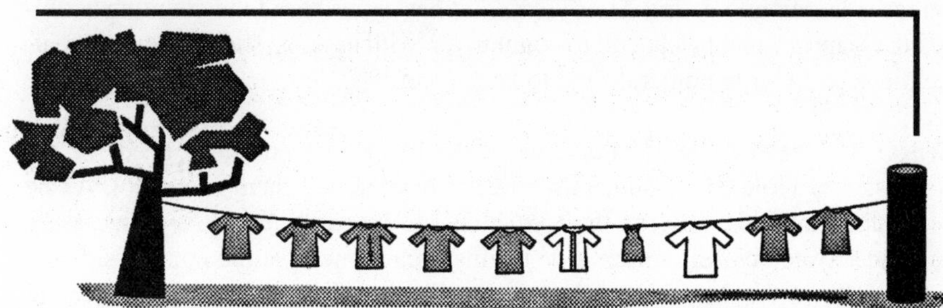

During your child's primary years in school, you can help build a solid foundation for his later study and use of mathematics. By guiding the child as he uses mathematics in the environment, you will become aware of the child's successes and difficulties. With your direction, his natural curiosity will lead to active learning situations that reinforce the skills and concepts introduced in the classroom. These experiences will increase the child's appreciation of mathematics as a tool for discovering answers and for communicating ideas to others.

Throughout this chapter, activities are described that will reinforce your child's use of the numbers to ten. These activities are designed to help the child realize the need to record numbers for communicating ideas to others and for his own use at a later time. This realization makes the act of writing seem less of a chore and more of a convenience.

Though textbooks vary as to when they present some topics, here is a list of topics most often included for kindergarten or early first grade.

Kindergarten or Grade One:

1) Counting and writing numbers to ten;
2) ordinal numbers: first to tenth (1 st to 1 Oth);
3) money: pennies, nickels, dimes;
4) measurement of length: inches, centimeters; and
5) addition and subtraction facts with sums to ten.

Use this list for reference and as a guide for determining the skills that need to be reinforced and practiced by your child. If you are in doubt, consult with the child's teacher, who may wish to check those items which the child needs special help with or those items that he has learned well enough to use in daily activities.

Counting and Writing Numerals to Ten

One of the most important steps in building a foundation for the study of mathematics is learning to count. After your child has experienced the counting activities in Chapter 1 and has begun to count to ten with ease, he should begin to realize the need to write numerals and to keep records.

ZERO TO NINE: Create situations in the home that present a need for the youngster to write these numerals after he has learned them in school. As he writes the numerals, observe the formation and be ready to give any necessary guidance. Watch closely to see if he begins each numeral at the point beside the X and follows the direction of the arrows. He must lift the pencil for a second stroke for the numerals 4 and 5.

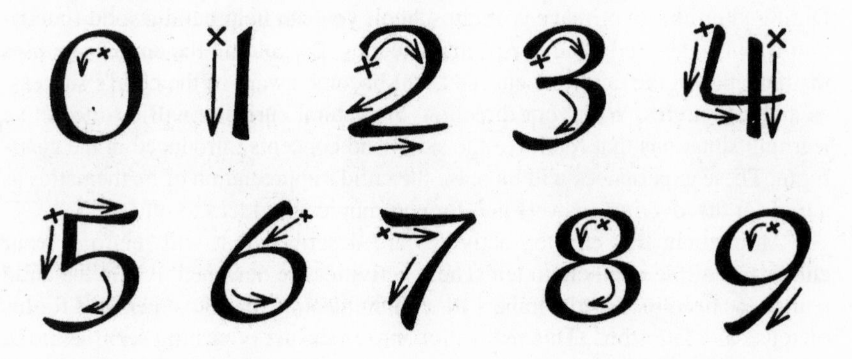

Involve your child in activities that require him to count zero to nine items and have the child record the numbers. These activities will help him develop an appreciation for the use of numbers and for the value of keeping records. At this level, keep all activities short to prevent the child from becoming tired.

Examples:

1. When you prepare a grocery list, give your child a picture list similar to the following. Ask him to count the pieces of fruit in the kitchen and record the number of each kind beside its picture so that you will know how many of each you need to buy.

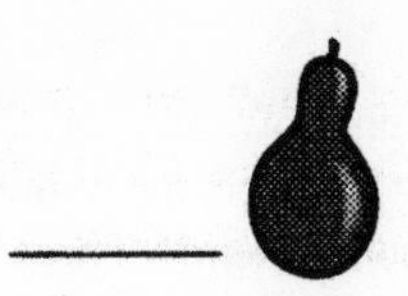

If you have a picture of one kind of fruit that is not in the kitchen, check to see if he records this by writing Q (zero). After he has completed the list, ask the child to read it to you.

"Four apples."

2. When you are sewing, ask your child to count the buttons of each color in a box. Give him a picture similar to the one shown here with each button a different color and have the child record the number of each.

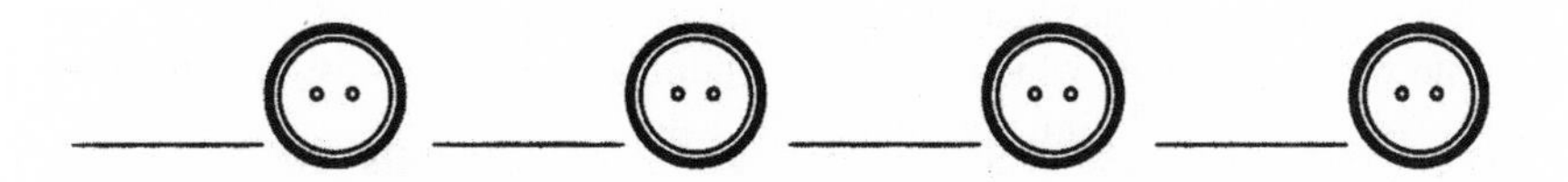

Again, encourage the child to read the completed list to you.

3. While woodworking, ask your child to record the number of nails of each size.

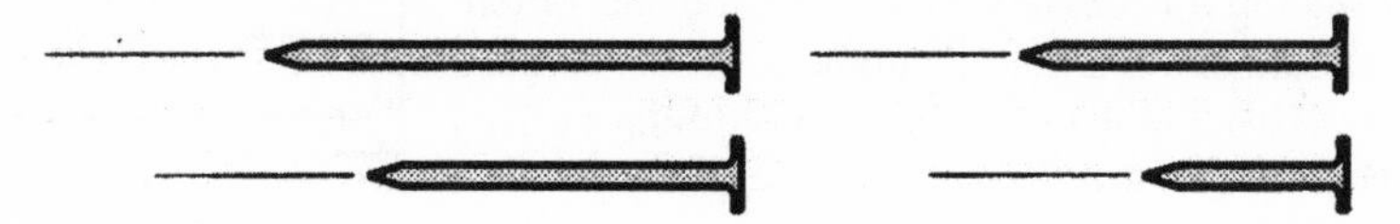

4. Children seem to enjoy any activity more if someone is involved with them. Say to your child: "Let's each write the number that tells:

a) Your age: ___;
b) the number of fingers on one hand: ___:
c) the number of chins you have: ___;
d) the number of eyes you have:___;
c) the number of comers on a triangle:. f) the number of comers on a square: _
g) the number of fingers I'm showing (Show six, seven, eight, and nine fingers): ___; and h) the number of elephants in the room: ___."

To develop readiness for addition and subtraction, encourage your child to begin counting with a number greater than one.

Examples:

1. Say: "I have three pennies in the bank. Let's count how many are in the bank as I put in some more. We will begin counting with three, the number already in the bank, and say each number as I put in another penny. Three-four, five, six, seven."
2. Say: "There are five boards in this stack. Count the number of boards as I put on some more. What number will you say first? (Five.) Continue count* ing as I place the boards."
3. Say: "There are two apples in this bag. Count the apples in the bag as I put in some more." Be sure he says the number two before you put any apples in the bag.
4. Say: "There are four people here for dinner and I see three more people com ing. How many people will be here for dinner?"

Provide other situations where the child will begin to count with a number greater than one. To find the answer to 3 + 2, the youngster will count "three -four, five" rather than "one, two, three - four, five."

TEN: After your child has learned to form all of the numerals. 0 to 9. he is ready to write the numeral 10 which is the first number that is expressed with place value. The place of the digit determines the value.

Example:

In the numeral 10, the one indicates one group of ten and the zero indicates no ones.

Give the child a chart like Fig. 1 and help him read words at the top if necessary. Give the child sets of ten or less to count and record the number of tens and ones in each set. The completed chart will be similar to that in Fig. 2.

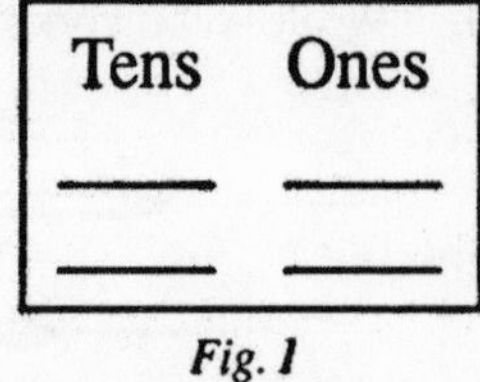

Fig. 1

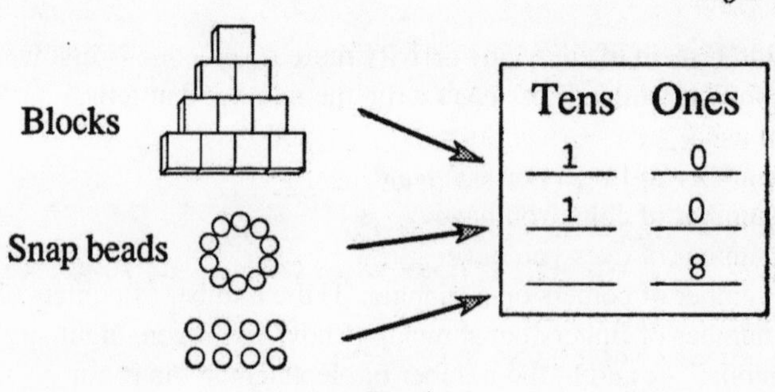

Fig. 2

Stress that any whole number less than ten requires only one digit so the space on the left is blank for those numbers. Since the numeral on the left tells the number of tens, no numeral will be put there for numbers zero to nine.

To provide practice for counting and writing numerals to ten, give the child a box of buttons (poker chips, marbles, or other small items) with ten or fewer of each color and a chart similar to the one below. Be sure that there are ten of some colors. Tell him that you need to know how many buttons there are of each color and want him to son the buttons by color and then count and record the number in each set. Show the child how to put each set of ten in a stack. Provide paper clip boxes or other small containers for each set of ten or have the child string each set of ten on a thread. Ask him to use his crayons to put a dot of the same color as the color of the button beside the numeral that tells how many. The completed chart will look something like this:

Ask the child to read the chart to you.

	Tens	Ones
(Black dot) ●	___	8
(Blue dot) ●	1	0
(Green dot) ○	___	5

"There are eight black buttons, ten blue buttons, and five green buttons."

After the child has been involved in activities similar to the one above, ask him to write the numeral that tells the number of toes he has; the number of fingers; the number in any other sets of ten that are available. Do not use a chart to specify the location of the numerals 1 and_Q, but stress that the 1 is on the left to tell how many sets of ten and the Q is on the right to tell how many ones there are.

COMPARISON: GREATER THAN (>) AND LESS THAN (<): Your child will be learning to compare number values by identifying the number that is greater than, or less than, another number. To make these comparisons in writing, he will need to use the signs: > for is greater than and < for is less than.

If your child has been involved in one-to-one matching activities described in Chapter 1. he is probably able to tell whether a number, 0 to 10, is greater or less than another number, 0 to 10. He may have difficulty identifying the sign that is to be used to complete a comparison sentence, so any clues you can give will be helpful.

Examples:

1. Show pictures of two sets of food treats and ask the child to draw a mouth between the two sets, opened toward the set with more. (The set he would like to eat). After he has drawn the open mouth, show him how to write and read the matching number sentence.

Draw:

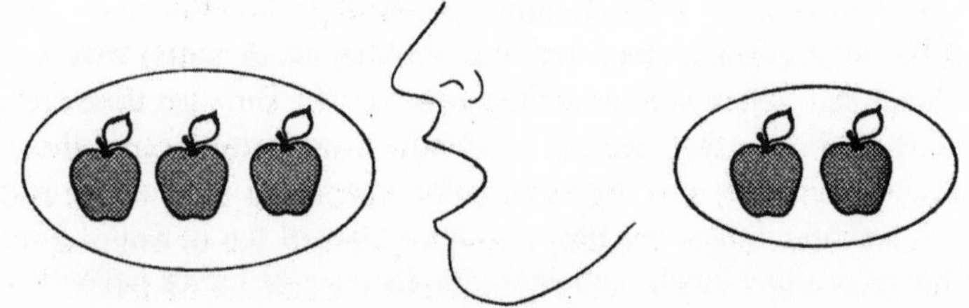

Write: 3>2
Read: "Three is greater than two"

2. Draw:

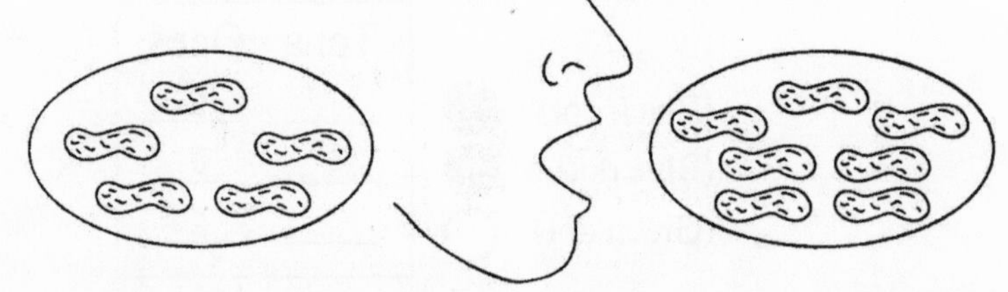

Write: 5<7
Read: "Five is less than seven"

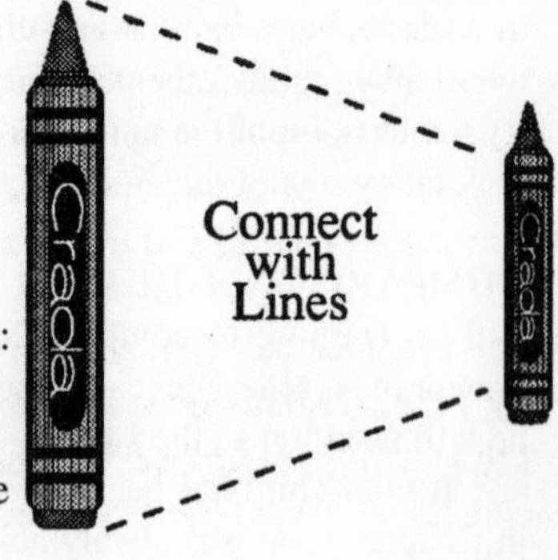

3. Draw two crayons. 3 inches and 1 inch long. Discuss the fact that lines connecting the ends resemble the greater than sign. (>) Write the following sentence:

   .3 inches ____ 1 inch.

   Encourage the child to measure the crayons and write the sign in the ____ that will make the sentence true. Ask him to read the sentence: "Three inches is greater than one inch."

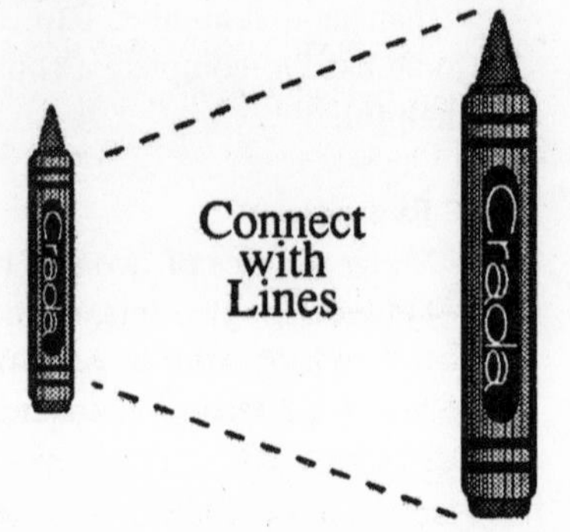

4. Follow the same procedure with the crayons reversed.

   1 inch < 3 inches

   "One inch is less than three inches."

5. Have each of the players throw a die (one of a pair of dice), in turn, and record the number or dots on top of his die. Have them compare the two numbers after each round.

Player l.		Player 2.
1	**<**	**3**

(Note: < is a sign for "is less than" and > is a sign for "is greater than.")

The player with the higher roll wins the round and gets one point. After nine rounds, the player with more points wins the game.

Ordinal Numbers 1st-10th

If your child has learned to count to ten. he is ready to identify the position of any object in its order from first to tenth. Help him learn to use these ordinal names by using activities similar to the following:

1. Give the order of participants for a game-Jumping rope, bouncing a ball- or other activity by using the ordinal names. Say to the participants: "Line up in this order. Mary, you will be first. Jean. you will be second so get behind Mary. Joe. you will be third so get behind Jean. Mary is number one. or first. Jean is number two, or second. Joe is number three, or third."
2. As you indicate a set of books lined up on a shelf or table, say to your child: "Please bring the fourth book from the left to me." If the child has difficulty following directions, show him how to count the books from the left to find the fourth one.
3. Give the child some crayons and a drawing similar to the following:

Say: "Color the second shirt from the tree green. Now color the third shirt from the tree yellow. Color the eighth shirt from the tree brown." In order to stress the point of origin for finding the ordinal position, change your directions to specify a position from the post. Say: "Color the second shirt from the post blue. Color the fourth shin from the post red. What color do you want the fifth shirt from the post? Color it" Continue with directions for coloring the other shirts.

4. Use the picture colored in Activity 3 to provide opportunities for the child to answer questions about the order. Ask such questions as: "Which shirt is green? Count from the tree. What is the color of the sixth shirt from the tree? What color is the third shirt from the post?"
5. Make ten tags and label each with an ordinal number from 1st to 10th. Pin a tag on each of ten or fewer people to tell each person's place in order for an activity such as receiving a treat or jumping rope.

1st	2nd	3rd	4th	5th
6th	7th	8th	9th	10th

6. Use the tags that you prepared for Activity 5 to label each participant in a race according to his place in the order of completion of the race. The per son that finishes first will get the label 1st pinned to his shin. The one that finishes second will get 2nd pinned on his shirt.

Money Pennies, Nickels, Dimes

There are advantages to be gained by reinforcing the value of pennies, nickels, and dimes after the child has developed an understanding of numbers one to ten. The use of these coins provides practical activities that involve computation and counting to higher numbers.

Great care must be taken when asking questions about sets of coins. Specify whether you are asking how many coins or the value of the coins. If you show a nickel and three pennies and ask how much, the answer can be four coins or it can be eight cents from the child's point of view. Explain that if you ask how many the answer will be 4 coins; if you ask how much, which is the question usually asked about money, the answer will be 8 cents.

Give the child a set of coins that includes pennies, nickels, and dimes. Label toys, food treats, or other attractive items with prices of 1¢, 2¢...10¢. Explain the cent sign (¢) as a short way to write cents. Ask him to give you enough money to buy an item that costs 1¢ to 4¢ . Then ask for enough money to buy an item that costs 5¢. If she shows 5 pennies, ask if she can use any one coin to buy the item. Explain that (he nickel has the same value as 5 pennies.

Follow the same procedure for having the child purchase items for \0f. After the child is able to identify the values of nickels and dimes, give him a nickel and three pennies. Ask how many cents this is worth. Encourage the child to begin with "five" and continue counting the pennies. "Six, seven, eight." Involve him in other exercises similar to this. Use one nickel and one to five pennies.

Label items with prices from 6¢ to lO¢. Provide nickels and pennies for each purchase. Let the child count the money he needs to buy each item. For items labeled 10¢, ask him to pay with pennies (10); nickels (2); and then dimes (1).

Measuring Length with Inches and Centimeters

Stir the child's curiosity by asking such questions as: "Which is longer? Shorter?" Watching you perform measuring activities around the home to find answers to these questions will also increase interest. After he has learned to use units of measure, ask questions similar to those above to motivate the child to count and compute.

Give the child items that are not over ten inches long (envelopes, pencils, fingers) and ask him to measure the length. To keep the activity as simple as possible, buy or make a ruler marked with only inches and half inches. Show the child how to place the end of the object at the end of the ruler and read the number nearest to the other end of the item.

Urge the child to practice her measuring skills by finding the answer to such questions as: "About how many inches long is your pencil? Your thumb? Each finger? The red pencil?"

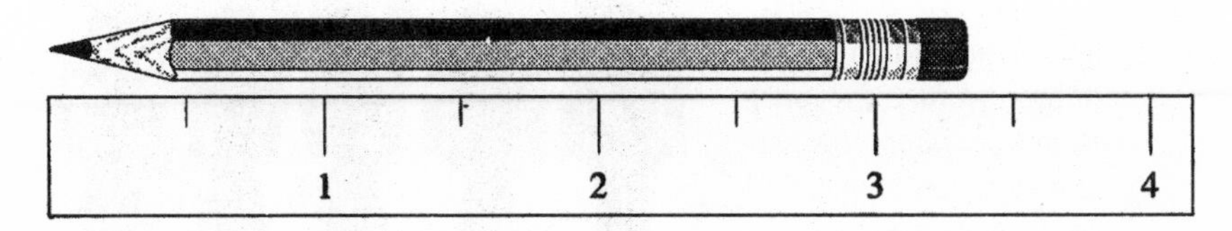

MEASURING WITH CENTIMETERS: To teach your child to measure with centimeters, use a metric ruler. Involve him in activities similar to those with inches.

Addition and Subtraction Facts with Sums to Ten

After your child has learned to count and use numbers to ten, he is ready for experiences thai require addition and subtraction with sums to ten. Of course, he needs to learn how to add and subtract, but it is just as important for him to team when to add or subtract. To develop skill with these processes, he should:

a) Experience situations that require addition or subtraction to solve real problems that affect the child's daily life;

b) learn to use the vocabulary and operation signs that are necessary to describe these processes orally and in writing; and

c) memorize the facts.

Provide the experiences named in steps a) and b) (solving real problems and writing number sentences) by asking the question "how many?" as you follow your daily routine. Make the problems as meaningful to the child as possible.

PROBLEM SOLVING; WITH ADDITION: It is important for the child to learn that when two or more sets are joined, the plus sign (+) is used to show that the numbers are added. Create problem solving situations that require your child to find sums to ten. Show him how to write the number sentences and use the names for the parts of each sentence when discussing it.

Addend plus addend equals sum.
3 + 2 = 5

or

3	Addend
±2	Addend
5	Sum

Some examples of activities that require addition are:

1. When planning a meal, say: and three women for lunch. (Use any numbers that will give a sum less than ten.) I need to know how many plates (knives, forks, or spoons) to put on the table. Can you find how many?" If necessary, draw a picture of two men and three women, write the number sentence below the picture, and tell the child this sentence is one way to record the number of people altogether.

   2 + 3 = 5

   Count the number of people and complete the sentence (2+3=5). Read the completed sentence. "Two plus three equals five." Explain that the sum (5) tells the number of people who will be coming for lunch and how many plates to set. At this point the child is not only learning that the sum of two and three is five, he is also learning to use the words plus and equals to read an addition number sentence (equation).

2. When picking flowers, say: "We can put two flowers in this vase and one flower in the other vase. How many flowers shall we pick?" After the child has made an effort to find out, draw a picture and write the number sentence below it.

   2 + 1 = _____

   Ask the child to read the number sentence and tell what number you should write on the blank.
3. When arranging toys, say: "There are two cars and two trucks in the box. How many people can have a toy if each person gets one? How many toys are there altogether?" Ask the child to write the number sentence, but provide as much help as he needs. Encourage him to read the sentence to you.

Continue to use every situation that you can create to encourage the child to find the sum of two numbers with sums of ten or less.

PROBLEM SOLVING WITH SUBTRACTION: It is important for the child to learn that when some members of a set are removed, the minus sign (-) is used to show that the number of members taken away is subtracted from the number of the original set. The result tells the number left. As you create problem solving experiences, keep the minuend (number of the original set) to ten or less. Stress that the order of the numbers is important. The number of the original set is on the left of the minus sign and the number to be removed is on the right of the sign. Create problem solving situations that require your child to find differences between numbers to ten. Show him how to write each number sentence and always use the names of the parts of the sentence when discussing it.

Minuend minus subtrahend equals difference.
7 - 5 = 2

or

7	Minuend
-5	Subtrahend
2	Difference

Some examples of problems involving subtraction are:

1. When preparing a snack, say: "We have five carrot sticks. If we eat two now, how many will be left?" If necessary, draw a picture, write the number

sentence below it, and explain that the number sentence shows one way to record that two carrot sticks are taken from five carrot sticks. Ask how many will be left and record three as the difference. Read the completed sentence. "Five minus two equals three."

2. When playing Store, say: "You have eight pennies. If you give five pennies to buy a pencil, how many pennies will you have left? Write the number sentence." (8 - 5 = _) After the child has written the sentence, let him take five pennies from eight pennies and write the difference on the __.

If a child is taught subtraction only as a "take away" process, he will be confused when he meets word problems that involve the comparison of two numbers-a problem that asks how many more or how many less. Some examples of situations that require the answers to those questions are:

1. When sharing a treat with some children, say: "There are five children and here are two pictures to color. How many more pictures do I need to get from the coloring book so that each child will have one?" Show two ways to write the number sentence: 2 + __ =5 or 5-2= __ . Encourage the child to write the number sentence and, if necessary, draw the picture and write the subtraction sentence under the picture.

   5 - 2 = _____

   Say: "There are three children left without a picture. Five minus two equals three. So I need to get three more pictures."

2. When counting money, say: "How many pennies do you have? (3) I have one penny. Who has less? How many less? How can we find out?" You may find it helpful to place her pennies in a row with yours below it and match one-to-one. Write the missing-addend sentence and the subtraction sentence and encourage her to write the difference.

   1 + __ =3
   3 - 1 =__

Continue to create situations that will encourage your child to use the process of subtraction to solve "take away" and comparison problems that affect his daily life.

PROBLEM SOLVING WITH ADDITION AND SUBTRACTION: After the child has used addition and subtraction sentences, give other problem situations and let him solve each by determining which type of number sentence to write. Stress that:

a) When sets are joined, or put together, a plus sign (+) is used and numbers are added, b) When sets are separated (taken away) or compared (How many more? How many less?), one number is subtracted from the other and a minus sign (-) is used.

Examples:

1. As you give your child some money, say: "You have three pennies. If I give you four more. how many will you have? Write a number sentence to (ell how many." (3 + 4 = _) Give him the four pennies and have your child count all of his pennies. Stress that since two sets were joined the plus sign is used and the numbers are added.
2. Say: "Now you have seven pennies. If you trade five for a nickel, how many pennies will you have left? Write the number sentence." (7 - 5 = _) If he writes a missing-addend sentence (5 + _ =7) tell him that is one way to write the sentence and ask him to write it another way. Stress that the first sentence he wrote is an addition sentence and (he second is a subtraction sentence. Encourage him to write the difference, then take the five pennies and have your child count the remaining ones to check his answer.
3. Say: "My blue crayon is three inches long. Yours is four inches long. How long are the two crayons when they are placed end to end?" (3 + 4 = _)
4. When preparing for a meal. say: "There are four chairs at the table. There will be six people for dinner. How many more chairs do you need to get? Write the number sentence." (6 - 4 = _)
5. Say: "I must drive 6 nails in the roof of the dog house and 4 nails in the bota tom. Which will have more nails, the roof or the bottom? How many more?" (6 - 4 = _)

The child will also be learning to write the addition and subtraction examples in vertical form 2 and 3.

±2 ±2

so encourage him to use both methods, vertical and horizontal, to find the sums and differences.

MEMORIZING THE FACTS: Now your child is probably ready (o memorize the addition and subtraction facts with sums to ten. This process may take a period of months, so be patient. During the primary years, your child needs to learn to give the following sums and differences accurately and rapidly:

0 +0	0 +1	1 +0	0 +2	1 +1	2 +0	0 +3	1 +2	2 +1
3 +0	0 +4	1 +3	2 +2	3 +1	4 +0	0 +5	1 +4	2 +3

3	4	5	0	1	2	3	4	5
+2	+1	+0	+6	+5	+4	+3	+2	+1
6	0	1	2	3	4	5	6	7
+0	+7	+6	+5	+4	+3	+2	+1	+0
0	1	2	3	4	5	6	7	8
+8	+7	+6	+5	+4	+3	+2	+1	+0
0	1	2	3	4	5	6	7	8
+9	+8	+7	+6	+5	+4	+3	+2	+1
9	0	1	2	3	4	5	6	7
+0	+10	+9	+8	+7	+6	+5	+4	+3
8	9	10						
+2	+1	+0						

— — — — —

0	1	1	2	2	2	3	3	3
−0	−0	−1	−0	−1	−2	−0	−1	−2
3	4	4	4	4	4	5	5	5
−3	−0	−1	−2	−3	−4	−0	−1	−2
5	5	5	6	6	6	6	6	6
−3	−4	−5	−0	−1	−2	−3	−4	−5
6	7	7	7	7	7	7	7	7
−6	−0	−1	−2	−3	−4	−5	−6	−7
8	8	8	8	8	8	8	8	8
−0	−1	−2	−3	−4	−5	−6	−7	−8
9	9	9	9	9	9	9	9	9
−0	−1	−2	−3	−4	−5	−6	−7	−8
9	10	10	10	10	10	10	10	10
−9	−0	−1	−2	−3	−4	−5	−6	−7
10	10	10						
−8	−9	−10						

Begin with the easiest facts and present only a few at a time for your child to practice until he has gained speed and accuracy.

ADDITION OF ZERO: Of the 66 addition facts, 21 have zero as an addend. These are simple to learn because zero plus any number equals that number. (0 + 6 = 6 ; 2 + 0 = 2)

SUMS WITH MATCHING ADDENDS. Each of 20 facts has the same addends as another fact (reversed in order), therefore it has the same sum. (1 + 2 = 3 and 2 + 1 = 3). If the addends are the same, the sums are the same.

2	1
±1	±2
3	3

After identifying the facts with zero as an addend and discovering the 20 pairs of matching facts, we can see there are only 5 more addition facts with sums to ten for us to learn. These are called the doubles facts. (1 + 1, 2 + 2, 3 + 3, etc.) These facts are comparatively easy for children to learn.

SUBTRACTION OF ZERO: Of the 66 subtraction facts, 11 have a subtrahend (number to be subtracted) of zero. Most children learn quickly that any number minus zero equals thai number. (6 - 0 = 6.2 - 0 = 2, etc.)

A DIFFERENCE OF ZERO: Of the 55 subtraction facts without zero as a subtrahend, 10 have zero as a difference. When a number is subtracted from itself, the difference is zero. (4 - 4 = 0,7 - 7 = 0, etc.)

MATCHING SUBTRAHEND AND DIFFERENCE: There arc 45 other facts. Of these, 20 have the subtrahend and difference reversed from 20 other facts. When the child has learned the first fact the matching fact should be easy.

$$\begin{array}{r} 5 \\ -2 \\ \hline 3 \end{array} \leftrightarrow \begin{array}{r} 5 \\ -3 \\ \hline 2 \end{array} \qquad \begin{array}{r} 7 \\ -3 \\ \hline 4 \end{array} \leftrightarrow \begin{array}{r} 7 \\ -4 \\ \hline 3 \end{array}$$

There arc 5 other subtraction facts with sums (top number) to ten for your child to learn.

$$\begin{array}{r} 2 \\ -1 \\ \hline \end{array} \quad \begin{array}{r} 4 \\ -2 \\ \hline \end{array} \quad \begin{array}{r} 6 \\ -3 \\ \hline \end{array} \quad \begin{array}{r} 8 \\ -4 \\ \hline \end{array} \quad \begin{array}{r} 10 \\ -5 \\ \hline \end{array}$$

FACT FAMILIES: Since subtraction is the "undoing" of addition, we can write a "fact family" for any pair of addends.

Example: For 3 + 2 = 5 the other members of the fact family arc:

2+3=5

5-3=2

5-2=3

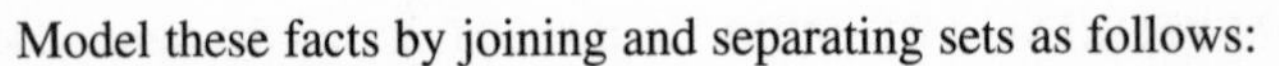

Model these facts by joining and separating sets as follows:

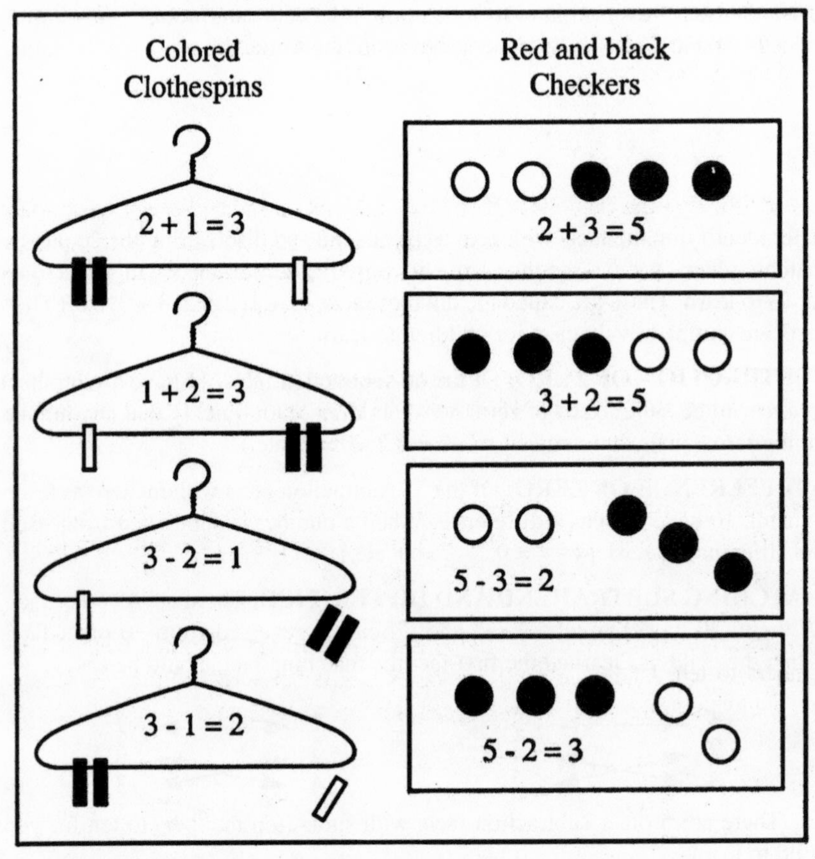

For some combinations, there are only two facts in the family.

Examples:

4+4=8	8-4=4	3+3=6
6-3=3	2+2=4	4-2-2

Learning these fact families may help your child memorize the facts. When he has learned one fact, he will also know the matching fact.

REMEMBER!

Speed and accuracy come after understanding.

Drill and practice are absolutely necessary.

It is important to keep the drill and practice as enjoyable as possible, but the child should be aware that he is working to memorize the facts (sums and differences).

To provide practice for the addition facts, choose any of the following from "Games," Chapter 11:

The Mystery Box, Games 2.4
Fill My Pockets, Game 1
Pay the Piper, Game 1
Facts Relay Race. Game 1
Tic Tac Toe
Match Up, Game 2
The Trail Game

To provide practice for the subtraction facts, choose any of the following from "Games," Chapter 11:

The Mystery Box, Games 3.5
Empty My Pockets
Pay the Piper, Game 2
Facts Relay Race, Game 3
Tic Tac Toe
Match Up, Game 3
The Trail Game

Note the sums and differences that the child gives slowly or incorrectly as he plays the games. Encourage the child to practice these facts by repeating the games listed above, using only the facts that he still needs to learn.

Summary

The focus of this chapter has been on helping your child develop skills in using numbers to ten. Understanding and memorization of addition and subtraction facts with sums to ten has been stressed. Though practice is necessary to develop speed and accuracy, the routine must be interspersed with real problems so the child will know why it is important to learn the process and when to use these new skills. Remember, while some problems and games may seem useless to adults, they are needed to prevent practice from resulting in boredom for your child.

3
USING NUMBERS TO 99

While your child is practicing addition and subtraction facts with sums to ten, he will also be learning to count and to read and write numerals to 99. As with the number ten. all of these numbers have place value involved. The value of each digit, ones or tens, is determined by its place in the numeral, on the right or left

Example: In the numeral 13 the 3 names three ones but in the numeral 31, the 3 names three tens or thirty.

This sounds simple to adults, but the primary child sometimes has great difficulty with the concept. You can help your child grasp the meaning of these numerals and thus build a better foundation for future mathematics by providing meaningful grouping, counting, and recording activities.

This chapter includes activities for the development and practical uses of these numbers. The following topics, usually included in first and second grade textbooks, will be developed.

Grades One and Two:

1. Counting and writing numbers to 99;
2. using the calendar,
3. fractions:1/^,1/), '/4 of a region;
4. addition and subtraction of two-digit numbers;
5. time on the hour.
6. addition and subtraction facts with sums to eighteen;
7. graphing; and
8. probability.

Continue to consult with the child's teacher if you have any questions about his readiness for any concept or skill.

Counting and Writing Numbers to 99

Tens	Ones

As with previous counting activities, suggest that the child help you with organizational tasks by counting and recording the number of items in various sets commonly found around your home. Include the numbers 10 to 99. Ask the child to group the items by tens, record the number of tens on the left, the number of ones on the right, and then ask the child to read the number to you. If he is just beginning to study these numbers, begin with ten and continue with the numbers in order. Stress that the number of tens is written on the left and the ones on the right.

Make a chart similar to the one on the right. Ask the child to record numbers from the following activities on this chart.

1. Give the child ten to 99 pennies. Have him count the pennies, stack them into sets of 10. and exchange each stack for a dime. Encourage the child to record the tens and ones on the chart and read the number that tells how many cents. Examples: 45¢ and 54¢; 83¢ and 38¢; 17¢ and 71¢.
2. Cut nine individual squares and one strip of ten squares from a sheet of graph paper or draw them as shown. Call the individual squares "unit-squares" and each strip of ten a "ten-strip."

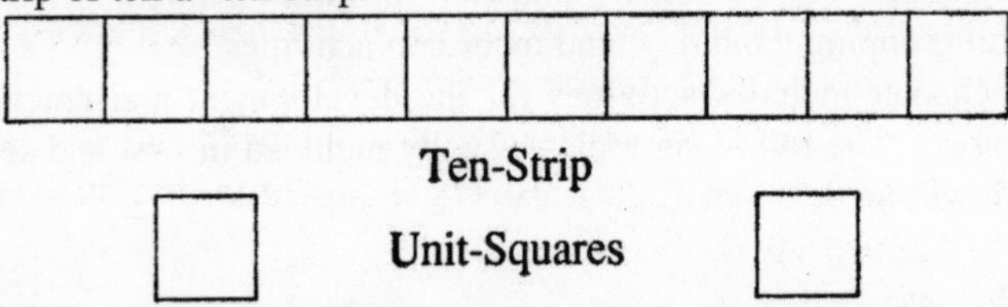

   Show one ten-strip and one unit-square and ask the child to name and write the number on the chart. (11) Model other numbers, 10 to 99, and ask the child to name and record each.
3. Show the child toothpicks, tongue depressors, or popsicle sticks bundled in sets of ten and some individual sticks. Ask him to record the number on the chart.
   Examples: 3 bundles, 5 sticks (35)
   5 bundles. 3 sticks (53)
4. Write a number from 10 to 99 and ask the child to model the number with any of the above materials.

Involve your child in activities for counting to 99 that do not require grouping

items by ten. **Examples:** (1) Counting marbles from a bag; (2) counting cars on the highway; and (3) counting roadside signs. After he is able to identify the number on the left as the number of tens and the number on the right as the number of ones, let the child write these numbers without the aid of the chart Have him locate page numbers, read numbers from the newspaper, read number labels, read dates on a calendar, or read temperatures on a thermometer.
Draw a 10 by 10 grid of 1 inch squares and help your child make a 0-99 square like the one shown below. Glue the square to a piece of cardboard and cover with clear contact shelf paper. You can mark on this with a crayon and wash the marks off later with most household cleansers (ammonia, glass cleaner, etc.). Hang the square in a prominent place in the home, a place where it will be visible when the two of you are having conversations and where the child will see it often. Use it now and save for other uses later.

0	1	2	3	4	5	6	7	8	9
10	11	12	13	14	15	16	17	18	19
20	21	22	23	24	25	26	27	28	29
30	31	32	33	34	35	36	37	38	39
40	41	42	43	44	45	46	47	48	49
50	51	52	53	54	55	56	57	58	59
60	61	62	63	64	65	66	67	68	69
70	71	72	73	74	75	76	77	78	79
80	81	82	83	84	85	86	87	88	89
90	91	92	93	94	95	96	97	98	99

After you have explained the meaning of rows (-") and columns^), ask questions about the numbers on the chart similar to the following:

1. How are all of the numbers in the third row from the top alike? (They all have two tens.)
2. How are the numbers in the fourth column from the left alike? (They all have three ones.)
3. Color red all of the numerals with 5 ones. What column are they in? (Sixth from the left.)
4. Color blue all of the numerals with 5 tens. Where are they? (In the sixth row from the top.)

To strengthen the child's ability to count to 99 and read and write these numbers, encourage him to play some of the following games from the "Games" chapter of the book: Name the Number, Play Store, Games 1 and 2; Show the Number.

USING THE CALENDAR: Now that your child is learning to read. write, and model numbers to 99, use these numbers in as many ways in his daily life as possible. The calendar provides an opportunity to improve the child's perspective of time in days and increase knowledge of the sequence of numbers.

Use the day's date and the day of the week often in conversations. Each morning, show the calendar and ask the child to tell the month, date and the day of the week. Give as much help as is necessary. Locate the day of the month and show the child how to move up the column to find the day of the week. Explain the abbreviations of the names of the days as a short way of writing the words. As you and your child look at a calendar, ask questions such as:

1. "Today is March 23. Grandma is coming in three days. What day will Grandma get here? What date?" Refer to any happening that the child anticipates.
2. "How many days are there until your birthday? Count (hem on the calendar."
3. "We're leaving for vacation on Friday. What is that date? We will be gone 14 days. What date will we get home?" In this example, if Friday's date is June 28 the vacation will continue into the next month-2 days in June and 12 days in July. Simple addition (28 + 14) will not provide the answer. Refer to the calendar if the child seems confused.

Your child will probably enjoy keeping a diary. At the end of each day he could record the date and day of the week on a sheet in a tablet or notebook. Suggest the child draw a simple picture of what he enjoyed most about the day. He may want to write a sentence or two that tells what he liked most.

Keeping this diary will lead the child to concentrate on all elements of the date each day-the day of the week, dale. month and year. It will stress the happiest part of the day and il will provide an interesting record for you and your child in the future.

NUMBERS BEFORE, AFTER, AND BETWEEN: To reinforce the counting order of the numbers to 99 for your child, involve him with questions about the numbers that come before, after, and between olhcr numbers.

Showing a calendar page. tell (he child that you will give clues and he must be a detective to guess the date thai you are describing.

Example:

1. Say: "We will go to the zoo (circus, a movie, the store) on the day with the number between 16 and 18. What is the number of that day?" (17) Keep the promise.
2. Say: "Grandma and Grandpa (any friend, relative, or acquaintance) will come on the day that comes after 23. What is the number of that day?" (24)
3. Say: "We will go on a picnic on (he day that comes immediately before 20. What is that dale?" (19)

When the child is able to answer questions like these by looking at the calendar, ask similar questions without letting the child look at the calendar. He can use the calendar to check his answers.

Play a detective game when locating page numbers in a book.

Example:

1. Say: "I am looking at a page between pages 49 and 51. What is the number of that page?"
2. Say: "The story begins on the page that comes after page 65. What is the number of that page? Find it."
3. Say: "I want you to find the page thai comes before page number 73. What is the page number? Find it."

COMPARISON OF TWO DIGIT NUMBERS: Before your child begins to write comparison sentences for two-digit numbers, he should be able to:

a. Compare one-digit numbers; and

b. use the sign for greater than (>) and less than (<).

Activities described in Chapter 2 will help the child learn these skills. It is slightly more difficult to learn to compare two-digit numbers than it is to learn to compare one-digit numbers. Your child must learn to compare the number of lens first. The number with more tens is greater, if the number of lens is equal, he must compare the number of ones.

46___36 (4 lens > 3 tens, so 46 is greater than (>) 36)

25___93 (2 tens < 9 tens. so 25 is less than (<) 93)

Remember, if he has difficulty with the signs use the activities from Chapter 2.

COUNTING BY TWOS: Tell your child that you know a fast way to count the pennies in his piggy bank. Demonstrate by putting 20 pennies in a pile. Move two pennies at a time and count by twos: two, four, six, eight, ten, etc.

Follow with activities similar to the following:

1. Place 30 to 50 pennies in a pile. Move two at a time as you and the child count by twos together.
2. Place 40 to 50 pennies in a pile. Let the child move the pennies and count by twos.
3. Provide less than 100 other small items for your child to count by twos.
4. Have the child count by twos the eyes, ears. hands or feet of the people pic tured in a clothes catalog.
5. On the 0 to 99 square prepared as described earlier in this chapter, have the child shade the numbers used when counting by twos.

COUNTING BY FIVES: Show counting by fives as a faster way to count than counting by ones or twos. Use activities similar to the following:

1. Place 30 pennies in stacks of five. Have the child count the pennies in some of the stacks to determine the number in each. Count the pennies by fives. Write the numerals. (5. 10, 15...30). Ask: "How are these numerals alike?" (They all have 5 or 0 in the ones place.)
2. Using the numerals in activity one, ask the child what number he thinks will come next when counting by fives. (35) Ask why. (Because the tens number stays the same as the previous number, 30, and 5 follows zero in the ones place in the pattern (0,5,10,15,20).
3. Talk about the fact that one nickel is worth five pennies. Place a number of nickels (less than 20) on the table. Guide the child as he counts how many pennies the nickels are worth. (Five, ten, fifteen.)
4. Have him count the fingers or toes by fives on pictures of people in books or catalogs.
5. Ask him to shade the numbers on the 0-99 square that are used when count ing by fives.

COUNTING BY TENS: Adjust the activities described for counting by fives to provide for counting by tens. For activity three use as many as nine dimes.

ODD AND EVEN NUMBERS: Use me following activities to reinforce these terms.

1. Have the child shade every other number (2,4,6.. .30) on a calendar with a favorite color. Have him identify these numbers as the numbers used when counting by twos. Tell the child that another name for these numbers is even

numbers. Ask if he knows the name for the numbers that are not shaded. (odd numbers)

2. Give your child 55 small items (pennies, checkers, poker chips) and ask him to use the items to build a model of each number, 1 to 10, by forming two rows as evenly as possible. Write EVEN on one sheet of paper and ODD on another sheet. Ask the child to place the models with two even rows under the word EVEN and record the numbers. Have him place the other models under the word ODD and record the numbers.

Even	Odd
2 4 6 8 10	1 3 5 7 9

Discuss how the models in each set are alike. (Each model of even numbers has two even rows. Each model of odd numbers has one more in one row than in the other).

3. At story time ask the child to tell whether page numbers are odd or even. Ask him to name the ones digits in the even numbers (0. 2.4. 6, 8) and the odd numbers (1. 3. 5, 7. 9). Point out that if the ones digit in any number is even, the number is even; if the ones digit is odd. the number is odd.
4. As you walk through your neighborhood, ask your child to identify house numbers as even or odd.
5. Suggest that he shade all of the even numbers on die 0 to 99 square.

After the child has learned to model and identify odd and even numbers, use this skill to reinforce addition facts introduced in Chapter 2.

Cut out the following models along the solid lines.

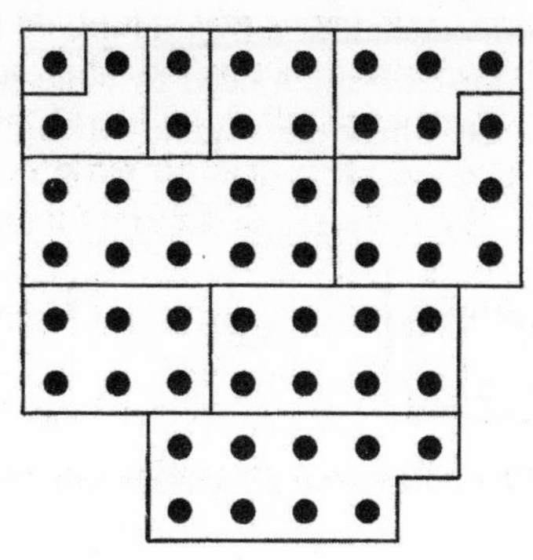

Use the models of odd and even numbers to show the patterns for adding numbers with sums to ten. Save the models to show addition with sums to 18 and multiplication facts at a later time.

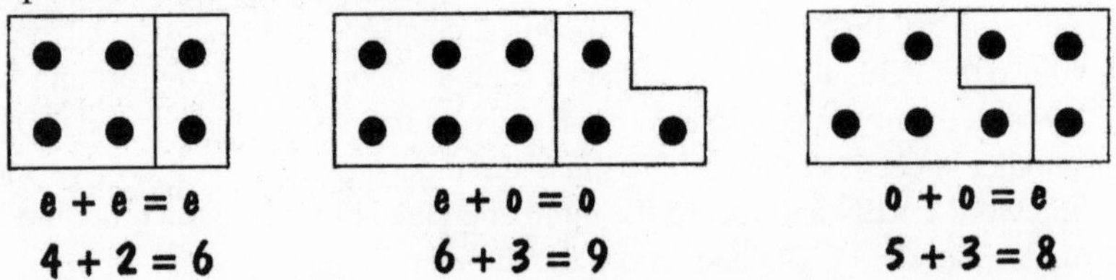

Show several examples of each category to prove that each statement is always true. These experiences stress that 6+7 must have a sum that is an odd number, therefore, 12 or 14 cannot be the correct sum.

To reinforce the concept, show addition examples and ask the child to circle the correct letter, "o" for odd and "e" for even.

8 + 5 < o e   9 + 4 < o e   6 + 4 < o e   3 + 5 < o e   7 + 9 < o e   4 + 8 < o e

Fractions
One Half, One Third, One Fourth

ONE HALF: Your child is probably aware of the term one half. Use a sharing activity to see if he knows that it means one of two equal pans. Give the child a sandwich and ask him to cut it or break it so that he can have one half and you can have one half. Remember - the one that cuts the sandwich gets second choice. After he has cut it, place one pan on top of the other to see if the two are the same size and, therefore, each pan is one half of the whole sandwich.

Continue with sharing activities that include the child finding one half of an object and using the term one half to tell about one of the two equal pans.

Remind your child that we often need to write number names in order to keep records for later use. Show the number name for one half. (1/2) Explain that:

One part out of two parts is $\frac{1}{2}$.	or	1 part out of 2 parts is $\frac{1}{2}$.

Show the child this number name on a measuring cup or on a half-cup measure:

Encouraging the child to fill the half-cup measure and empty it into the one-cup measure and talk about how much is filled. Repeat, and talk about how many of the half-cup measures it took to fill the one-cup measure.

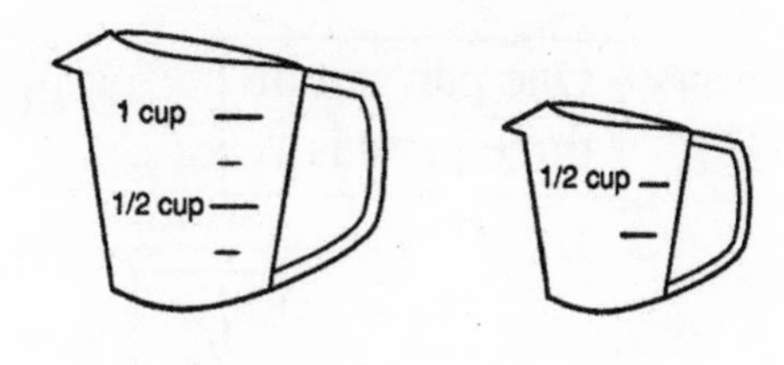

Show the child the number name 1/2 in recipes as you are cooking or baking. Discuss the fact that someone that you have never seen was able to tell you how to bake a cake or cook a casserole because he knew how to write words and number names; that you are able to do the cooking because you can read the number names and words.

Use a sheet of 8 1/2 inch by 11 inch paper to show the child one way to fold the paper in halves.

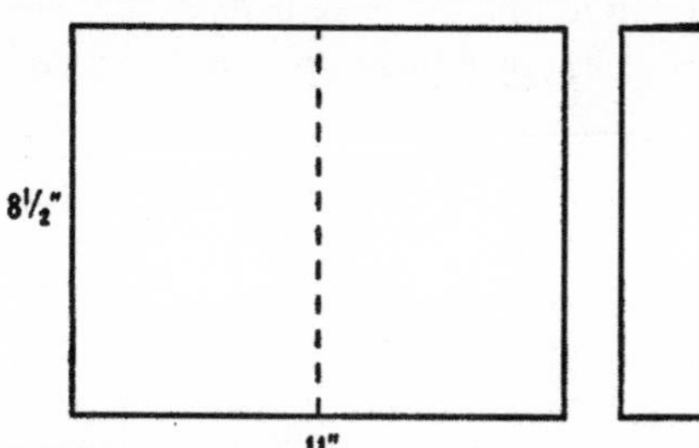

Give him another sheet and ask her to find another way to fold it into halves.
Cut two squares, 5 inches by 5 inches, from a sheet of paper and ask the child to fold each to show halves a different way. If she doesn't make a diagonal fold, show him.

On each folded paper, write directions for coloring:

1. Color ½ red.
Color ½ blue.

2. Color ½ green.
Color ½ yellow.

ONE THIRD: Use the sharing activity described for learning about one half to introduce one third. Let the child share something that is shaped like a rectangle (cracker or sheet of paper) between three people, including himself. Encourage the child to make the pieces as close to the same size as possible, but do not expect perfection. Circle shapes will be entirely too difficult for him to cut into thirds. Explain that:

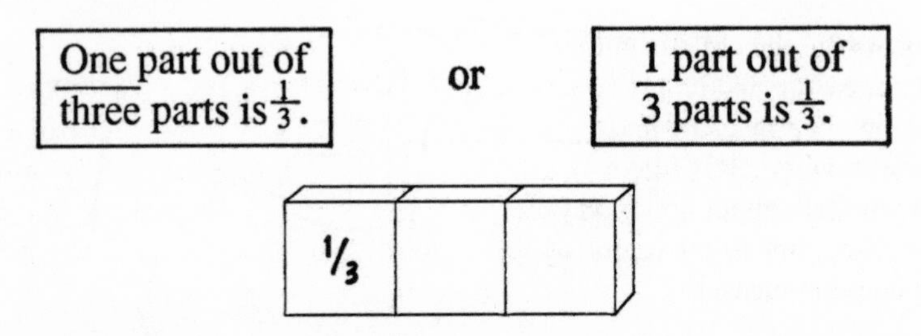

Use the measuring cups and recipes, as with one half, to reinforce the meaning and the number name of one third.

Adjust the paper folding and coloring activities described for learning one half. Do not expect the child to do the folding by him or herself because it is rather difficult to find the three parts of the same size. Using a rectangular piece of paper, 3 inches by 6 inches, you can use a ruler to find the following folds:

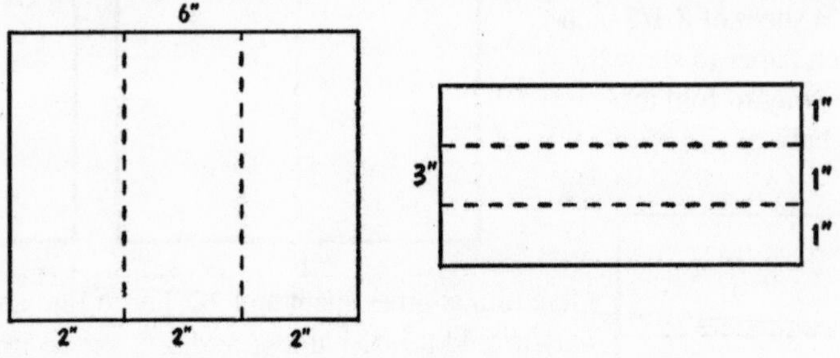

ONE FOURTH: If necessary, help the youngster divide round, square, and rectangular objects between himself and three other people (4), each person getting one fourth. Show the child the number name for one fourth (1/4) and explain as follows:

One part out of four parts is $\frac{1}{4}$.	or	1 part out of 4 parts is $\frac{1}{4}$.

For reinforcement of the number names and the importance of being able to record with number names, continue to use the measuring cups and recipes as you did for the fraction one half. Encourage the child to read a simple recipe to you as you cook or to follow a simple recipe as you read it to the child.

Use the paper folding and coloring activities as you did for halves and thirds.

ONE HALF, ONE THIRD, ONE FOURTH: After the child leams to identify these fractions, provide unfolded rectangles of paper with one direction written on each and observe his approach to following the directions. Display the finished papers in the home.

Color $\frac{1}{2}$ green.	Color $\frac{1}{4}$ yellow.	Color $\frac{1}{3}$ blue.

When working on a wood project, say: "I need to cut this board into halves (thirds, fourths). Where shall I cut it?" Observe the response to see if the child has a good perception of these fractional parts.

Addition and Subtraction Two-Digit Numbers without Regrouping

Adding and subtracting two-digit numbers (10 to 99) without regrouping (borrowing or carrying) should be easy for the child if he has learned addition and subtraction facts with sums to ten. Present some situations that require these processes and see if he can use them to find the answers.

Examples:

1. "I have 25< and you have 32?. How much do we have together?"
2. "We have 57< and the toy you want costs 89?. How much more do we need?"

If he has difficulty with these processes, use activities similar to the following. Be sure there are no more than nine ones or tens so that he will not have to regroup.

1. Show two dimes and three pennies and a set of five pennies. Ask the child to record the number in each set and find how many there are altogether. Remind him that when sets are joined the numbers are added. Stress that ones are added to ones and tens are added to tens.

Dimes Tens	Pennies Ones
2	3
+	5
2	8

2. Show a set of 34 and a set of 15 snap beads (paper clips, safety pins). Ask the child how many there are altogether. Stress that ones are added to ones and tens are added to tens.

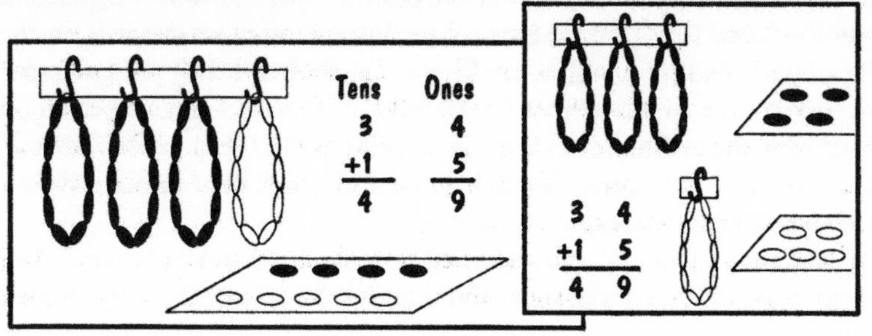

3. With dimes and pennies, show 47¢ and ask the child to record and show the value of the money that will be left if six pennies are taken away. Pennies are taken from pennies and ones are subtracted from ones.

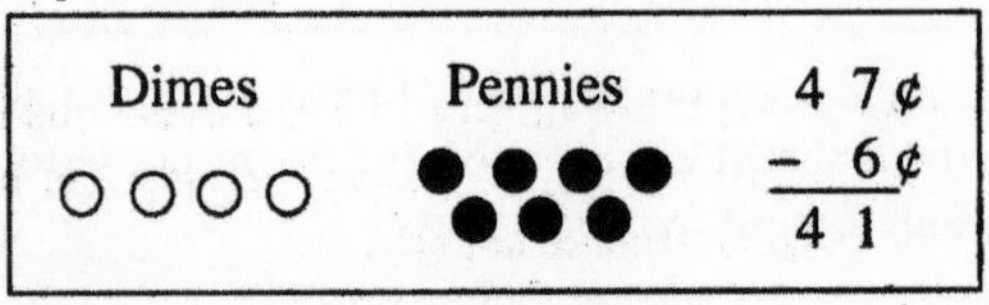

Dimes	Pennies	4 7¢
○○○○	●●●● ●●●	− 6¢
		4 1

After he has subtracted, encourage the child to take six pennies away and check his answer.

4. Show 47 paper clips (4 strings of ten and 7 loose clips) and ask how many will be left if 23 are taken away. After the youngster subtracts, ask him to remove the 23 clips to check the answer.

Stress: The number to be taken away in subtraction is the bottom number (-43) or the second number (75 - 41 = _). Ask questions about measurements given in inches to reinforce addition and subtraction and to motivate the child to use these processes.

Examples:

1. Ask him to measure two pieces of ribbon that you or he will use for a project. Ask for the total number of inches after he has found the two measures. (Addition)
2. Ask: "How much longer is the red ribbon than the blue ribbon?" (Subtraction)

Time on the Hour

Your child will be learning to tell time on non-digital clocks; therefore, it is important for these clocks to be available in the home. Since this process is a little more complicated than telling time on a digital clock, you may prefer to have your child's lirst watch be non-digital.

Use a real or toy clock or watch to show the movement of the long and short hands. Write some times on paper (1:00. 6:00, 12:00) and show the placement of the hands on the clock face for each of the times. Stress that on the hour. the long hand (minute hand) is on 12 and the short hand (hour hand) points to the hour. Write some other limes (7:00, 9:00, 10:00) on paper and encourage the child to show these times on the clock. Display some times on the clock and ask the child to read the times. A paper plate with cardboard hands held in place with a brad makes a good model clock.

Discuss activities that you and your child do at certain times each day. Help the child draw a clock face, show the time on (he clock, and draw a picture of the activity.

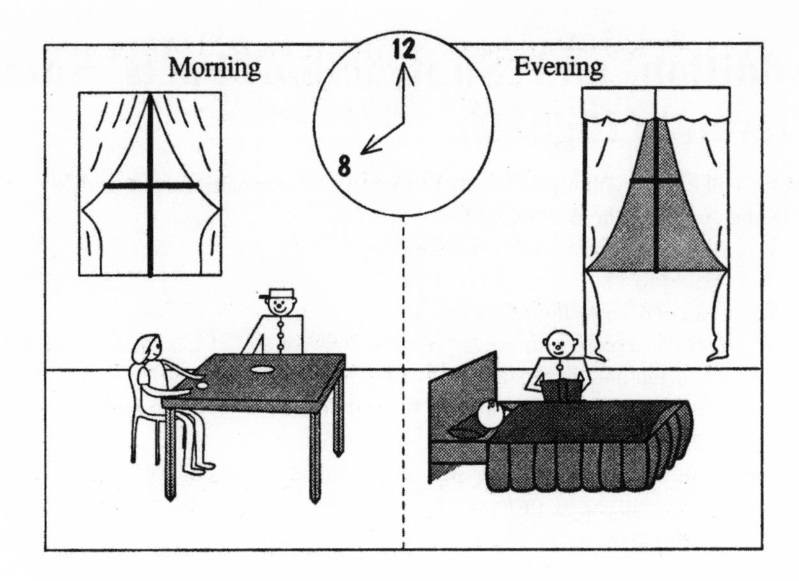

Talk about the times of favorite TV shows. Encourage the youngster to check the clock and read the time when the show begins.

When your child wants to visit a friend, talk about the time that he should come home. Let the child read the time on the clock. Be sure it is on the hour, possibly 2:00 o'clock. Say, "It is now 2:00 o'clock. In one hour it will be 3:00 o'clock. What lime will it be in two hours?" If necessary, suggest that he use the demonstration clock, move the hands around the face two times, and determine that in two hours it will be 4:00 o'clock and time to come home.

Use other situations of interest to the child to have him tell the time in a specified number of hours.

Examples:

1. "It is now 1:00 o'clock. Daddy (Mommy) will be home in four hours. What time will it be then?"
2. "The movie begins at 7:00 o'clock and is two hours long. What time will it be over?"
3. "It is now 3:00 o'clock. Dinner will be ready in three hours. What time will we eat dinner?"
4. "It is now 10:00 o'clock. We will leave for the movies in three hours. What time will we leave?" Watch this one closely. Does he add three, saying "13:00 o'clock"? Use a demonstration clock. Encourage the child to turn the hands to show that when three hours pass it will be 1:00 o'clock. Always be alert to situations when the elapsed hours will pass 12:00 o'clock.

   If you have digital clocks or watches in your home, show the times on the hour and encourage your child to read the time. This will help him learn to write time correctly. (6:00 o'clock; 12:00 o'clock)

Addition and Subtraction Facts Sums Eleven to Eighteen

Before you expect your child to lean) addition and subtraction facts with sums 11 to 18, be sure that she is able to:

a. Add and subtract with sums to ten;
b. count to 18;
c. read and write numerals to 18:
d. identify a teens number as "ten and some ones." (1 ten and 6 ones equals 16; 1 ten and 4 ones equals 14; 1 ten and 9 ones equals 19. etc.); and
e. use the vocabulary and signs to read and write addition and subtraction exercises.

Since our number system is based on ten, the processes of counting and computation are simplified by grouping by ten. The counting activities on the previous page have stressed the patterns of ones and tens that are repealed over and over. These patterns make the following sums and differences quite simple. Yet it is necessary that the child learn them at this stage.

$10+1=11$	$10+2=12$	$10+3=13$	$10+4=14$	
$\begin{array}{r}10\\+5\\\hline 15\end{array}$	$\begin{array}{r}10\\+6\\\hline 16\end{array}$	$\begin{array}{r}10\\+7\\\hline 17\end{array}$	$\begin{array}{r}10\\+8\\\hline 18\end{array}$	$\begin{array}{r}10\\+9\\\hline 19\end{array}$
$11-1=10$	$12-2=10$	$13-3=10$	$14-4=10$	
$\begin{array}{r}15\\-5\\\hline 10\end{array}$	$\begin{array}{r}16\\-6\\\hline 10\end{array}$	$\begin{array}{r}17\\-7\\\hline 10\end{array}$	$\begin{array}{r}18\\-8\\\hline 10\end{array}$	$\begin{array}{r}19\\-9\\\hline 10\end{array}$

If your child has difficulty giving these sums and differences, model some of the problems and guide him as he records the results.

Examples:

1. Show a ring of ten snap beads (paper clips, safety pins, freezer bag twisters, etc.) and encourage the child to write the number. Use the chart, if necessary. Show three more beads and ask him to write the number of that set and the sign that tells that the sets will be joined. (+)

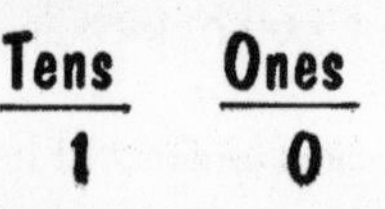

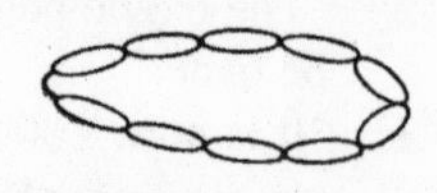

See that he places the 3 on the right side because it represents three ones. Next, ask the child write the sum. Urge him to join the sets to check the answer.

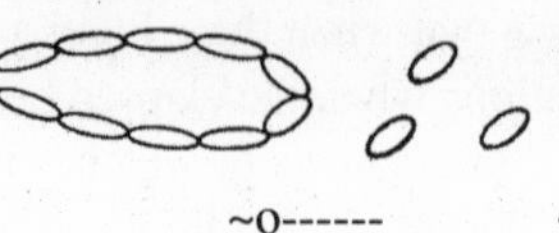

~o------

Tens	Ones
1	0
+	3

2. Show a set of seventeen beads—a ring of ten and seven individual beads—and ask the child to record the number. Ask how many will be left if he takes seven away and suggest that he write the sign and numbers that tell what happens. The seven must be written on the right because it names seven ones that are taken away from the ones, with the ten left as a unit.

Tens	Ones
1	7
–	7
1	0

If the child is able to give the above sums and differences rapidly and accurately, he is probably ready to learn the addition and subtraction facts with sums to 18.

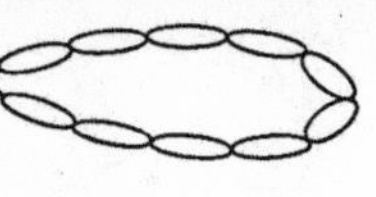

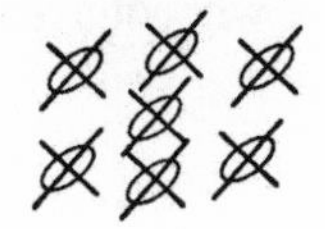

ADDITION FACTS:

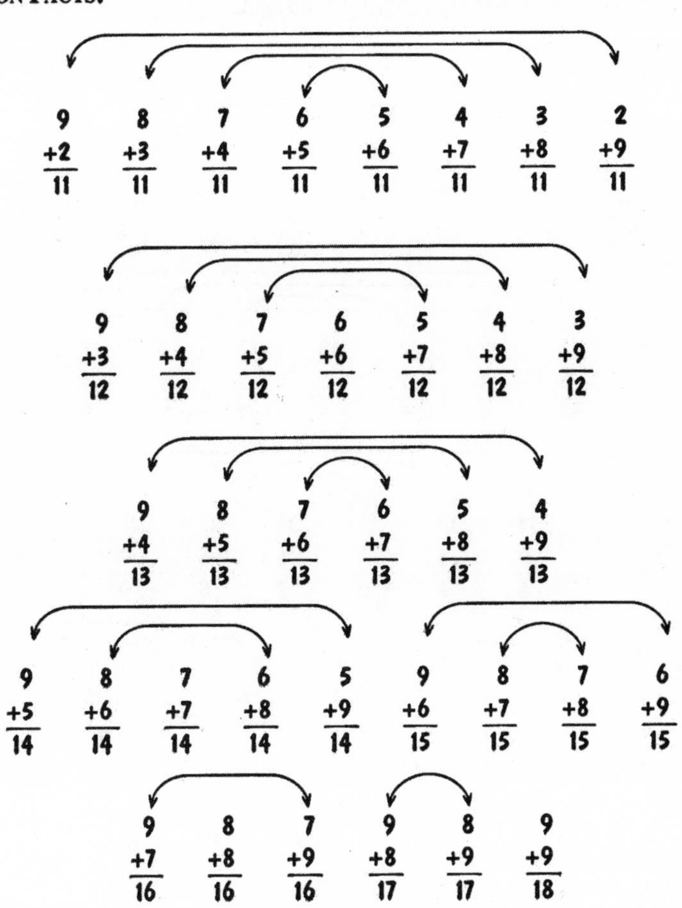

9 8 7 6 5 4 3 2
+2 +3 +4 +5 +6 +7 +8 +9
11 11 11 11 11 11 11 11
9 8 7 6 5 4 3
+3 +4 +5 +6 +7 +8 +9
12 12 12 12 12 12 12
9 8 7 6 5 4
+4 +5 +6 +7 +8 +9
13 13 13 13 13 13
9 8 7 6 5 9 8 7 6
+5 +6 +7 +8 +9 +6 +7 +8 +9
14 14 14 14 14 15 15 15 15
9 8 7 9 8 9
+7 +8 +9 +8 +9 +9
16 16 16 17 17 18

Of the 36 facts above, 16 have a matching fact. (Follow the arrows.) The addends are the same but are in a different order, so the sums are the same.

$$\begin{array}{r}8\\+7\\\hline 15\end{array} \longleftrightarrow \begin{array}{r}7\\+8\\\hline 15\end{array} \qquad \begin{array}{r}9\\+2\\\hline 11\end{array} \longleftrightarrow \begin{array}{r}2\\+9\\\hline 11\end{array}$$

Stress: "When you learn 8 + 7 = 15, you also know 7 + 8 = 15." Four facts are without a matching fact:

$$\begin{array}{r}6\\+6\\\hline 12\end{array} \quad \begin{array}{r}7\\+7\\\hline 14\end{array} \quad \begin{array}{r}8\\+8\\\hline 16\end{array} \quad \begin{array}{r}9\\+9\\\hline 18\end{array}$$

The 16 facts with a matching fact and the four facts without a matching fact make only 20 addition facts for the child to learn.

SUBTRACTION FACTS:

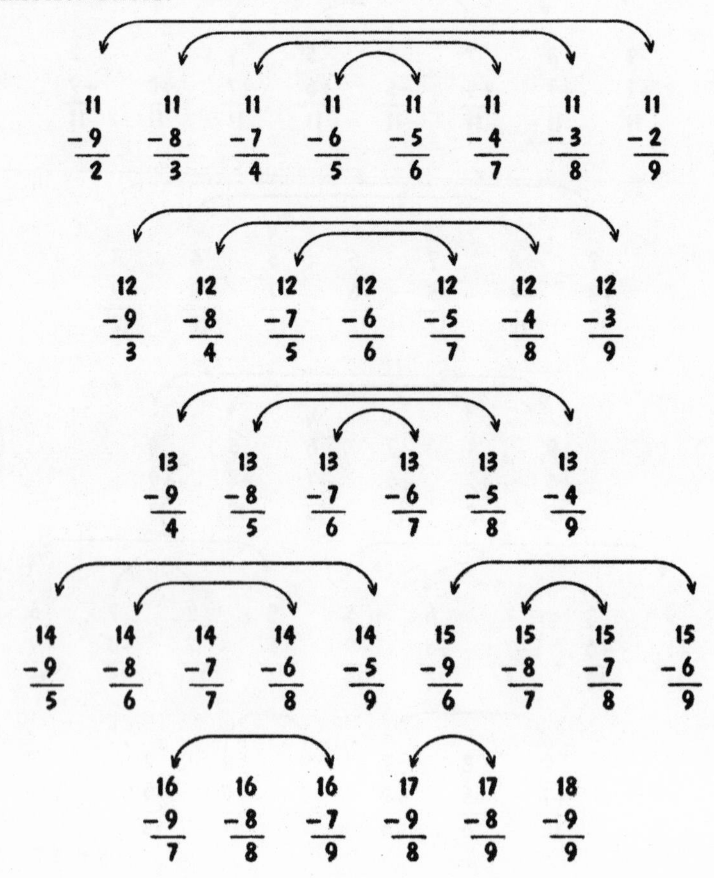

Again, 16 subtraction facts have matching facts; the subtrahend and difference are reversed. Stress: "When you learn $12-3=9$, you also know $12-9=3$."

$$\begin{array}{r} 12 \\ -3 \\ \hline 9 \end{array} \quad \longleftrightarrow \quad \begin{array}{r} 12 \\ -9 \\ \hline 3 \end{array}$$

There are four facts without a matching fact, so there are only 20 differences for the child to learn.

ADDITION AND SUBTRACTION FACTS: Create as many "real life" addition and subtraction problems with your daily activities as possible. In the beginning, use items that the child can easily bundle or group by tens.

Examples:

1. Say: "There are nine pink straws and seven blue straws in the box. Can you write an example on the chart that will help us find how many straws there are altogether? Will you join the sets or separate them? (Join.) Will you add or subtract the numbers?" (Add.)

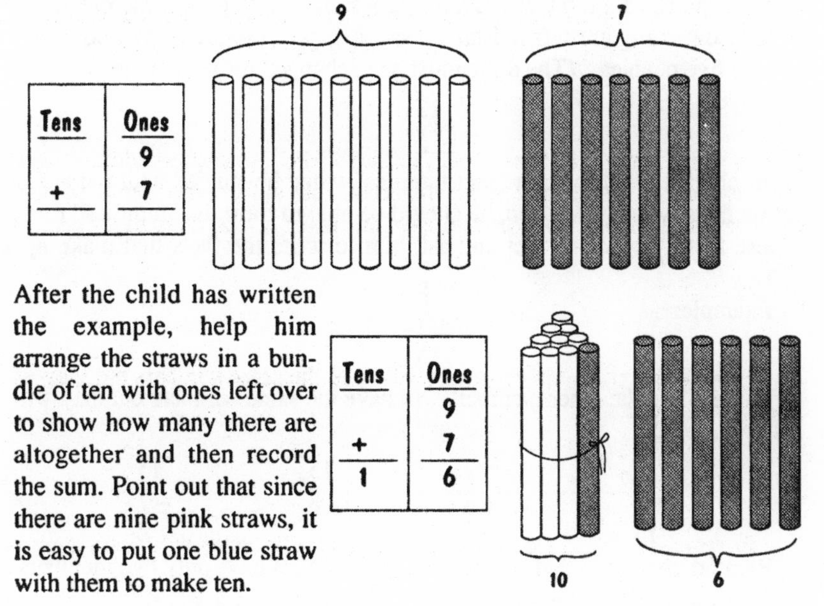

After the child has written the example, help him arrange the straws in a bundle of ten with ones left over to show how many there are altogether and then record the sum. Point out that since there are nine pink straws, it is easy to put one blue straw with them to make ten.

2. Show a ring of ten snap beads and six individual beads and ask how many there are. (Sixteen.) Say: "I want you to bring me nine of the beads. Can you write an example on this chart that will help us find how many will be left? Will you join or separate the sets? (Separate.) Will you add or subtract the numbers?" (Subtract.)

After he has written the example, talk about the fact that he

Tens	Ones
1	6
–	9

must unbundle the set of ten before nine beads can be taken away.

Tens	Ones
1	6
	9
	7

16

9

Stress that seven beads are left and he should write the seven on the right under Ones.

Show eight pennies and seven pennies. Have the child record the number to find how many there are altogether. Have him exchange the ten pennies for a dime and write the sum.

$$\begin{array}{r} 8 \\ +7 \\ \hline 15 \end{array}$$

Give the child a dime and three pennies. Say: "Write the example that will tell how many are left if I take away five pennies. Help me find a way to take five pennies." (The dime must be exchanged for ten pennies.)

$$\begin{array}{r} 13 \\ -5 \\ \hline \end{array}$$

your child learned to form "fact families" when he studied addition and sub-ction facts with sums to ten, he should be able to form the "families" for each these facts. Show one fact and ask him to write the facts that make up the nily.

Example:

$$\begin{array}{r} 8 \\ +7 \\ \hline 15 \end{array}$$

Explain that the other members will have the same numbers but they will be different order.

$$\begin{array}{r} 8 \\ +7 \\ \hline 15 \end{array} \qquad \begin{array}{r} 7 \\ +8 \\ \hline 15 \end{array} \qquad \begin{array}{r} 15 \\ -8 \\ \hline 7 \end{array} \qquad \begin{array}{r} 15 \\ -7 \\ \hline 8 \end{array}$$

Remind your child that some of the fact families have only two members.

$6+6=12$	$9+9=18$	$7+7=14$	$8+8=16$
$12-6=12$	$18-9=9$	$14-7=7$	$16-8=8$

Review the patterns for adding odd and even numbers modeled earlier in s chapter.

Even + Even = Even	Even + Odd = Odd	Odd + Odd = Even
$8+6=14$	$8+7=15$	$9+7=16$

To provide practice for the addition facts, choose games from the list below described in "Games," Chapter 11.

Pay the Piper, Game 1	Match Up, Game 2
Facts Relay Race,Game2	Tic Tac Toe
The Trail Game	Secret Sum
Secret Rule	Back to Back, Game 1

To provide practice for the subtraction facts, choose games from the list below that are described in "Games," Chapter 11.

The Trail Game	Tic Tac Toe
Facts Relay Race, Game 3	Match Up, Game 3
Secret Rule	Pay the Piper, Game 2
Back to Back, Game 1	

Graphing

A graph presents numerical facts in picture form so that they are easier to understand; therefore, your child will be learning to complete, read, and interpret simple graphs.

Draw graphs or use graph paper to make them similar to the following. Then look for situations in the child's life that will encourage him to graph information.

Suggested Activities:

1. Make a graph similar to Fig. 1. Place a pile of animal crackers on the table and ask the child to color a square beside the correct animal name for each cracker.

Bear							
Leopard							
Elephant							
Monkey							
Zebra							
Others							

Fig. 1 Horizontal Bar Graph

The completed graph will look something like Fig. 2.

Bear	■	■	■	■	■		
Leopard	■						
Elephant	■	■	■				
Monkey	■	■	■	■	■	■	
Zebra	■	■					
Others	■	■	■				

Fig. 2

After the graph is completed ask such questions as:

a. How many bears are there? leopards? elephants? (Count the colored squares in each row.)

b. There are the most of which animal? (If he cannot answer, ask which bar is longest and discuss the fact that since more squares are colored in that bar than any other, there are more of that animal than any other.)

c. There are the fewest of which animal? (Which bar is shortest?)

d. Are there more elephants or leopards?

2. Prepare a graph similar to Fig. 3. Have the child color 1 square for each hour (or half hour) that he watches TV.

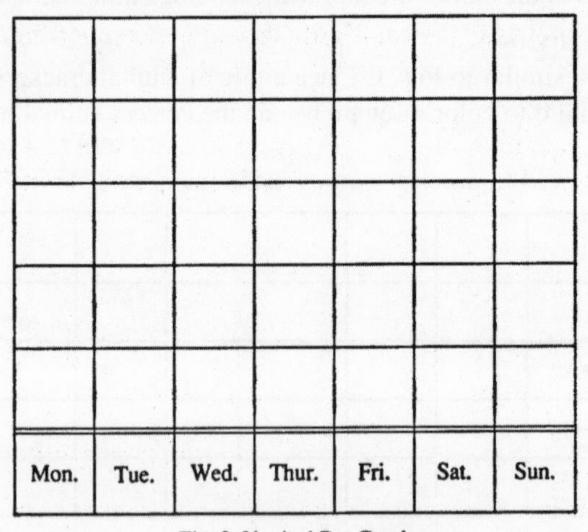

Fig. 3. Vertical Bar Graph

At the end of the week. ask: "On which day did you watch the most TV? The least? Was there a day that you did not watch TV? How many hours did you watch TV on Saturday?"

3. To encourage the child to practice measuring and graphing, ask him to graph the length of each family member's right foot. Watch your child's measuring techniques to see if he needs assistance. A page of quarter- inch

graph paper will be ideal for this activity. Let one square represent one inch of length.

Probability

In this modem age, so many of our life's decisions are based on probability. By giving your child an understanding of this important concept, you will help him make wiser choices. The study of probability helps us determine the chances an event has of occurring.

Examples:

1. If a person tosses a coin, only one of two sides-heads or tails-can show when it lands, so the probability for either side is / out of 2 or '/S
2. If a die with numerals 1 to 6, one numeral on each side. is rolled, the proba bility of any number showing is / out of 6 or 1/6,.
3. If there are 3 socks in a drawer-2 red and 1 blue-the probability of blindly pulling out a red sock is twice as great as that for a blue sock: red sock: 2 out of 3 or 2/3; blue sock: 1 out of 3 or 1/3.

Help your child develop an intuitive background for probability by engaging in activities similar to the following. Encourage your child to record the results of activities 2 through 7 on graphs to help him interpret the information and "guess" what will happen next

1. Display two boxes, each containing 5 checkers (buttons, cards, socks, or any items that are alike except for color). Write A on one box and B on the other. In Box A, place 4 red checkers and 1 black checker and in Box B. place 2 red checkers and 3 black checkers. Say: "I will give you a penny for your bank if you draw a red checker from one of the boxes without looking as you draw. From which box do you want to draw? Why?" Hopefully he will choose Box A because more of the checkers are red. Let the child draw several times, returning the checkers to the box after each draw.
2. Show a graph like Fig. 1 and suggest that the two of you take turns drawing checkers from Box A and record each draw on the graph by coloring a square in the proper column. Return the checker to the box after each draw. After approximately 15 draws, discuss which color was drawn more times (red) and why. Draw another graph and follow the same procedure with Box B. Compare the two graphs and ask: "From which box did you draw more black checkers? Why? More red checkers? Why?
3. Place 1 red and 1 black checker in a box. Tell the child that there Pg-1. are two checkers in the box and that they are either red or black. Show a

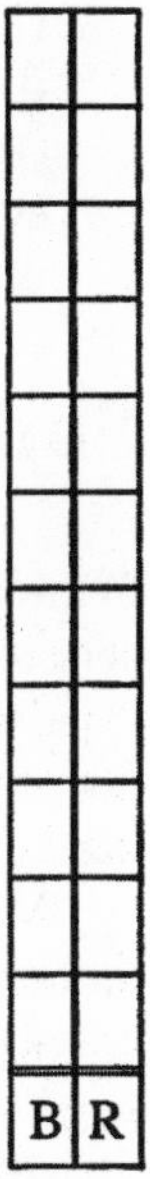

B	R

Fig. 1.

graph similar to Fig. 1. Ask him to draw a checker, record the color on the graph, and return the checker to the box. Encourage him to continue this sev eral times and then to guess how many of the checkers are red and how many are blue. Then show him the checkers.

4. Place 2 red checkers and 1 black checker in the box and follow the same procedure as in activity 3.
5. Place 1 red checker and 4 black checkers in the box and follow the same pro cedure as in activity 3.
6. Show a die and ask the child to guess which number will be on top when you roll the die. He will probably realize that the chances are equal for all sides. Suggest that he roll the die 20 to 30 times and record the result on a graph similar to Fig. 2. After the rolls, ask:"Did all numbers show about the same number of times?"
7. Show the following graph and two dice. Suggest that the child roll the dice several times and, after each roll. color a square above the sum of the two numbers that show.
   Example: 6 and 3 show so he should color a square above 9. After about 50 rolls ask: "Which sum was rolled the most times? (Probably 7.) Which sums were rolled the least number of times? (Probably 2 and 12.)"

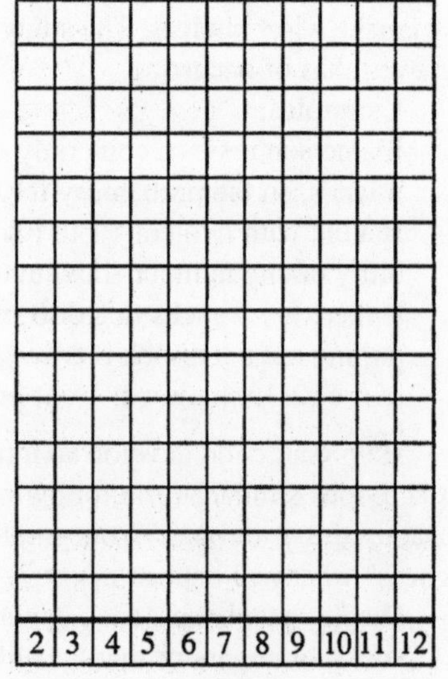

Fig. 2.

Summary

In this chapter your child has been introduced to the place value of digits in numbers 10 to 99. The practical applications of these numbers have been stressed by the use of the calendar, clock, graphs, probability, and activities that require addition and subtraction of 2-digit numbers without regrouping (carrying and borrowing).

Many activities have been suggested to provide practice of addition and subtraction facts with sums to 18. Continue with these activities until the child has memorized the facts and can give the sums and differences with speed and accuracy.

Practical uses of fractional numbers one half, one third, and one fourth have been suggested. Look for other uses of these numbers and integrate them into your child's activities at home.

4
USING NUMBERS TO 999

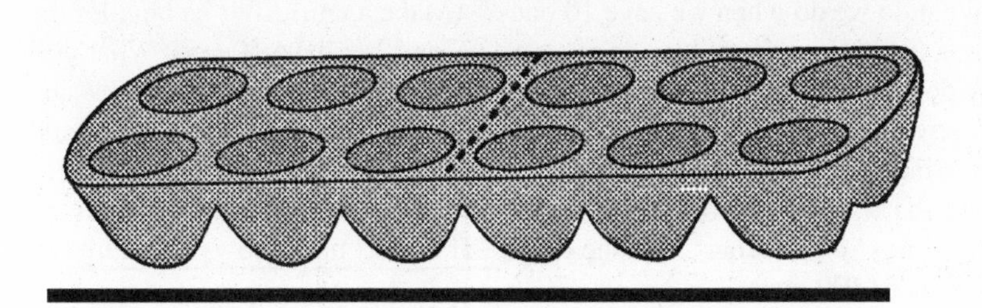

As your child's mathematical skills increase, the opportunities for him to apply these skills in regular activities also increase. He should use the skills learned as often as possible in daily life to maintain them.

The activities in this chapter are designed to reinforce the skills (listed below) that are usually taught in second or third grades.

Grades Two or Three:

1. Counting and writing numerals to 999;
2. money: quarters, half dollars, and dollars;
3. addition and subtraction of three-digit numbers without regrouping:
4. fractions: halves, thirds, fourths of regions and sets;
5. multiplication and division through the five facts;
6. time: five minute intervals; half hour,
7. liquid measures: cups, pints, quarts, gallons; liters; and
8. addition and subtraction of two-digit numbers with regrouping.

Textbooks vary with the sequence of skills, so frequent conferences with your child's teacher are of value.

Counting and Writing Numerals to 999

Your child will need to group ten sets of ten into a unit and record the number another place to the left as he leams to count past 99. To reinforce the place value of the hundreds place, use these activities:

1. Show nine bundles of ten sticks and nine ones. (tongue depressors, popsicle sticks, toothpicks, etc.) Ask the child to count the number of tens and ones and record the number.
   Place one more stick with the 9 ones and ask how many sticks there are. Observe to see if the child bundles the 10 ones to make 1 ten. If not, ask: "What do we do when we have 10 ones?" (Make a ten.) After he bundles the ones, ask: "How many lens do we have? (Ten.) We have 10 tens. What do we do when we have ten? (We bundle them.) So we'll bundle the 10 lens and what will we call that number? Yes, one hundred. (Use a large rubber band or string.) One bundle of 10 tens is one hundred. We can record the 10 tens on a chart. The name for the third place on (he chart is hundreds." Write "hundreds" on the chart. Tell the child: "The chart now shows 1 hundred or 10 tens or 100 ones."

Hundreds	Tens	Ones
1	0	0

   Continue the activity showing other sets of hundreds, tens. and ones from 100 to 999. Have the child record and read each number. Example: Show two bundles of one hundred (10 bun dles of ten in each large bundle), five bundles of ten, and three individual sticks and ask the child to record the number on the chart and read the num ber. 'Two hundred fifty-three." Note: there is no "and" in the number names unless there is a decimal point or a fraction. Correct: One hundred four teen. Incorrect: One hundred and fourteen.

Hundreds	Tens	Ones
2	5	3

2. Cut 10 individual squares (unit-squares), 10 strips of ten (ten-strips), and 9 ten-by-ten squares (hundred-squares) from graph paper.

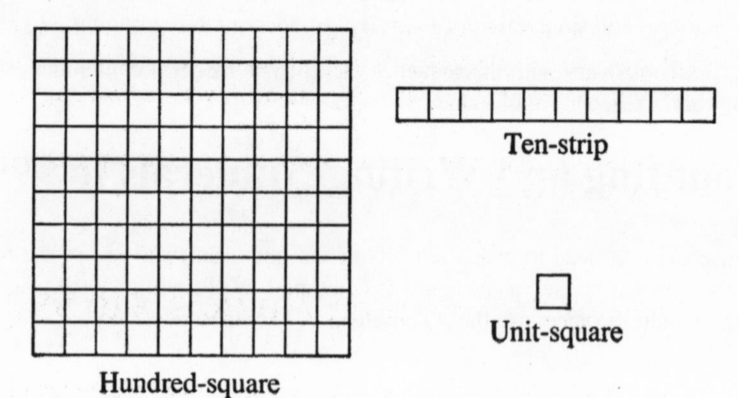

Explain that in this activity he must always trade 10 unit-squares for a ten-strip; and 10 ten-strips for a hundred-square. Give 9 unit-squares and 9

ten-strips to the child and have him record the number (99). Give the child one more unit-square and ask if he should make any trades. (Ten unit-squares for a ten-strip and 10 ten-strips for a hundred-square) Ask the child to record the number (100). Show three hundred-squares and have the child record the number (300). Model other numbers between 100 and 999; ask him to record and read each number.

3. Encourage the child to find other items to use to model the numbers from 100 to 999 (buttons on strings; bottle caps in stacks of ten with 10 stacks of ten put in a box to represent one hundred, etc.)

To develop the child's ability to estimate and to provide practice with counting, play an estimating game. Show a jar containing 100 to 999 pennies or other small items, and have members of the family guess the number of items in the jar. Let each contestant count the items to check the estimates. Encourage them to put the items in sets ot ten to make recounting easier. Give a small prize (treat from the kitchen, coin, holiday from a regular chore) to the person that guesses the number nearest the number of items.

QUARTERS AND HALF DOLLARS: Show your child a quarter and explain that the value is 25e, the same as 25 pennies. Suggest that he count out sets of 25 pennies from the money in the child's bank to trade to you for quarters. Ask if he can trade some nickels for a quarter. He can count the value of nickels by fives to 25 and count the number of nickels to determine that the value of five nickels is the same as the value of one quarter. Encourage the child to find other sets of coins to trade for a quarter (2 dimes and 1 nickel, 3 nickels and 1 dime).

After the child has become acquainted with the value of the quarter, provide opportunities to make purchases and use coins that include the quarter. Follow the same activities to acquaint the child with the half dollar.

Play the games Store, Game 2 and Buy or Sell from Chapter 11.

Allow the child to purchase items that cost 51¢ to 74¢. When he pays with a quarter and a half dollar, urge the child to name the amount of change due. Label items with the price shown as pan of a dollar ($0.75) and also with the cent sign (75¢).

Let the child have the opportunity to make purchases outside the home that involve the use of both coins.

DOLLARS: Your child may have been using dollar bills to make purchases. It is important that he knows its value in relation to other coins. Say: "I want to trade you a dollar bill for some dimes. How many dimes will you give me? The ten dimes are worth how many cents? If you trade a dollar for pennies, how many will you get? Yes. one hundred."

Follow the same questioning for the number of nickels, quarters, and half dollars in exchange for a dollar.

Show that the value of one dollar may be written as 100¢ but the more common way is to write it with a dollar sign, as $1.00. Write other amounts through $9.99. Have the child read and model each with bills and coins. Stress that dollars are written to the left of (he period (.) called the decimal point and cents to the right, as $ 1.01.

Using a catalog or newspaper ad, show the child prices that include dollars and cents. Read the price and help the child model (he amount with dollar bills and coins.

Refer to the Money chapter for other activities with money. Play Fewest Coins from Chapter 11.

Addition and Subtraction 3-Digit Numbers without Regrouping

When your child has learned the structure of three-digit numbers, 100 to 999, he should have little difficulty adding and subtracting these numbers without regrouping (carrying and borrowing). The child's teacher will probably establish the fact that it is customary to work from right to left, adding or subtracting ones, then tens, then hundreds. Observe closely as your child works at home to see if this sequence is followed. If not, consult the teacher to see if this is the pattern that has been established.

If the child has difficulty with these processes, adding and subtracting, model the exercises with any of the materials used for modeling the numbers in the section on counting and writing numerals to 999 in this chapter. Be sure that no regrouping is required at this time. Encourage the child to "act out" the process as indicated by the sign. + or -. He will join sets when the plus sign is used and separate sets when the minus sign is used.

Examples:

a) Addition

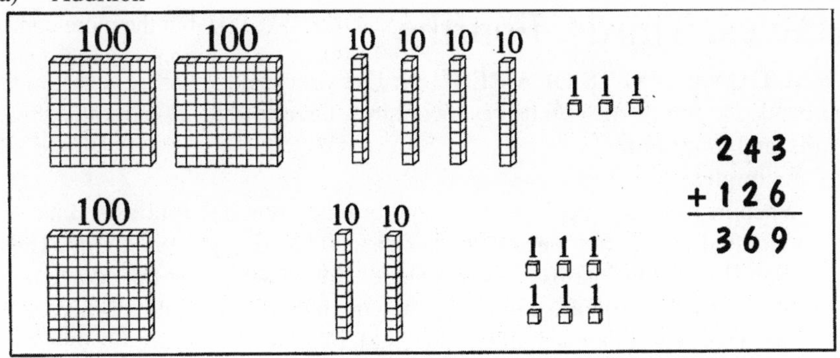

b) Subtraction

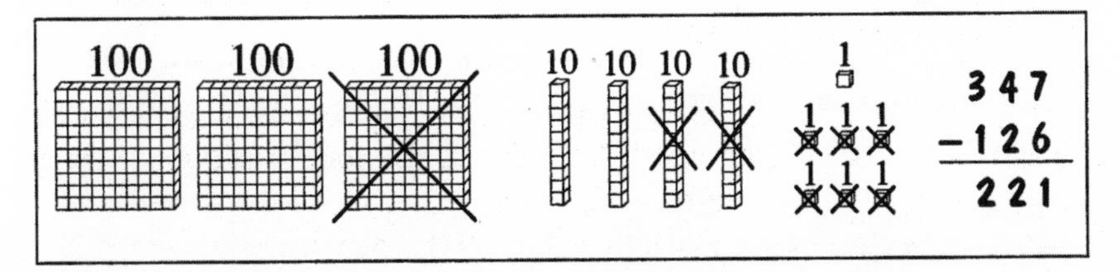

As you create situations around the home that require the child to add and subtract these numbers, have no regrouping (carrying and borrowing) involved. It is important to include situations that require sets or measures (inches) to be compared as well as joined and separated. The child will learn that the process of subtraction also answers the questions: How many more? Less? How much longer? Shorter?

Present activities similar to the following:

1. Tell the number of marbles in each of two bags and ask the child to find how many there are altogether without counting the marbles. (Join; add)
2. After telling the number of marbles in each bag, ask the child to find how many more there are in one than the other. (Compare; subtract)
3. Ask the child to find the total cost of two items listed in a catalog. (Join; add)
4. Ask him to find how much more one item costs than the other. (Compare; subtract)
5. Ask the child to find the total length of two pieces of rope or hose in inches. (Join; add)
6. After he has counted the pennies in his bank. ask how many will be left after he has bought some item listed in a catalog. (Separate; subtract)

Fractions
Halves, Thirds, Fourths

FRACTIONAL PARTS OF A REGION: Use your child's sharing activities to introduce the following: two halves, two thirds, three fourths, two fourths, three thirds, and four fourths.

Examples:

1. When he has cut a pizza, piece of fruit or other treat into fourths and has identified each part as one fourth, ask the child to give you one fourth. Then ask: "How many fourths are left? (Three) How much of the treat do you get?" (Three fourths) Ask him to show you the numeral name for this frac tion. If he doesn't know it, show the child. (%)
2. After he has divided a treat and shared it evenly between himself and two other people, ask: "How much did you give away? (Two thirds) How much did you keep?" (One third)

Coloring activities will reinforce the written names for these fractions. Show circles and rectangles divided into halves, thirds, and fourths and give written directions such as the following:

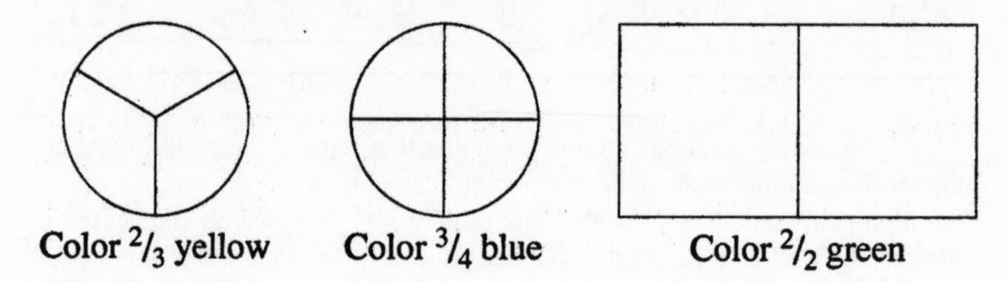
Color $^2/_3$ yellow
Color $^3/_4$ blue
Color $^2/_2$ green

Show the child the number names % and \ in recipes. Encourage him to watch as you follow the directions using these fractional names. Ask him to assist you by filling a measuring cup % full; \ full. Guide him as he follows simple recipes to prepare food. Stress: 2 parts out of 3 equal parts = ?5.

ONE HALF, ONE THIRD AND ONE FOURTH OF A NUMBER: You can simplify the process of finding the fractional part of a number by having the child find the fractional part of a set. Use activities similar to the following that model finding one half, one third, and one fourth of a set and a number.

1. Place six poker chips on the table: three red and three blue. Say: "How many poker chips are there? (6) One half of the chips are red and one half arc blue. How many are red? How many are blue? One half of six are how many?"(3)
2. Place six grapes on the table. Say: "You can have one half of the grapes. How many can you have? (3) Yes, one half of six is how many?" (3)

3. Place six pencils on the table. Say: "You can have one third of the pencils. How many can you have? (2) How many is one third of 6?" (2) If the child has difficulty, show him how to place the pencils in three sets with the same number in each set Ask how many are in one of the sets.(2)

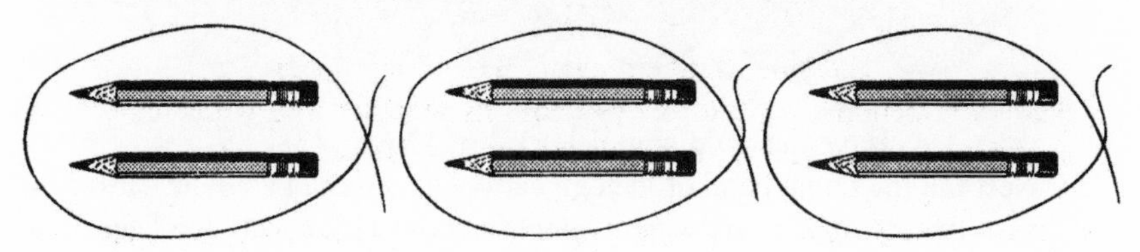

4. Give the child 8 pennies. Tell the child to find how much one fourth of 8 is and to give you one fourth of the pennies. If he has difficulty, demonstrate as in activity 3.
5. Use 4 egg canons to find one half, one third, and one fourth of twelve. Show two canons: 1 white and 1 blue. Guide the child as he cuts the blue canon to show halves. If you don't have colored canons, the child can paint or color them.

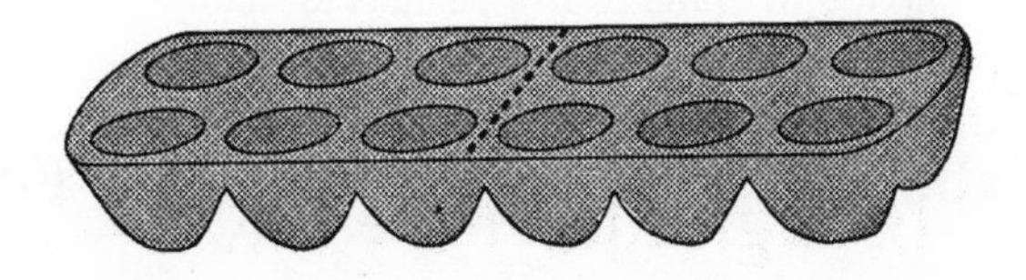

Place the two equal pans (halves of the carton) into a whole carton. With the child watching, remove one pan and ask: "How much of a whole carton is this?" (One half)
Place the two halves back into a canon and ask: "How many eggs will a carton hold? (12) How many eggs will one half of a carton hold? (If neces sary, take one half out and let the child count.) How much is one half of 12?" (6) Write the following sentence and ask the child to complete it

1/2 of 12 = __ (6)

Display a pink egg canon. Encourage the child to cut it to show thirds. Proceed as with halves.

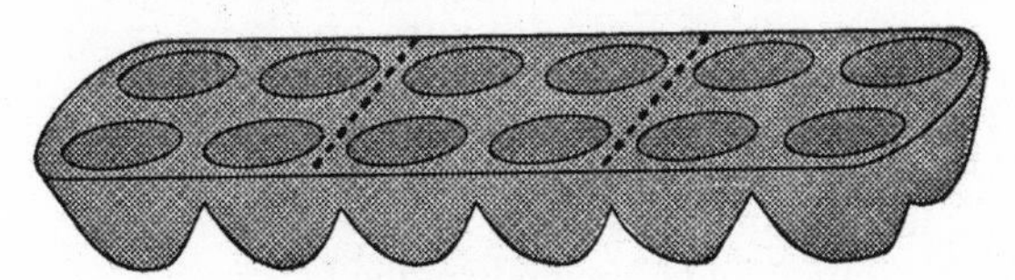

Have the child complete: 1/3 of 12 = __(4)

Follow the same procedure with a yellow canon to show four equal parts. (Fourths) Have the child cut it into four equal pans. Proceed as with halves.

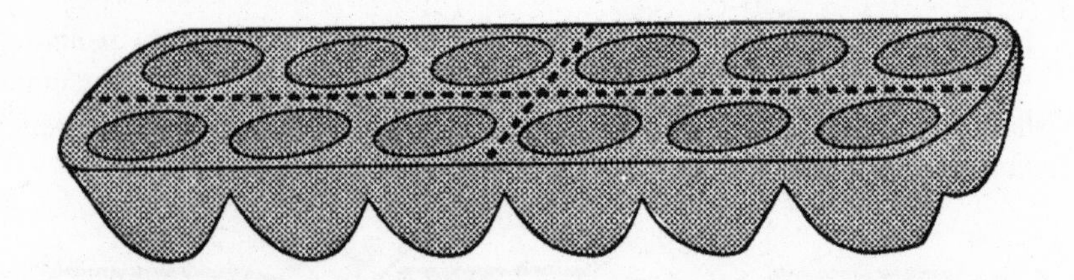

Have the child complete: 1/4 of 12 = __ (3)

Nest the fractional parts of the egg cartons together in a whole carton and save for activities to show equivalent fractions (1/2 = 2/4) at a later time.

Continue to use the terms one half, one third, and one fourth often when conversing with the child.

Examples:

1. Say: "There are 8 apples in the basket. Will you bring half of the apples to me? One half of eight equals how many?"
2. Say: "Six pieces of fruit are in the box. You may have one third of the pieces. One third of six equals how many?"
3. Say: 'Take one fourth of the twelve peanuts to share with your friends. One fourth of twelve equals how many?"

Multiplication and Division Through the "Five Facts"

Usually children anticipate learning multiplication and division with great enthusiasm. They have heard older brothers and sisters or friends talk about learning the "times tables," or the "five facts," and to some. it has become almost a symbol of growing up. This anticipation makes introducing the facts much more interesting.

If the child has learned that we use addition of numbers to describe joining sets. he can learn that multiplication of numbers describes joining equivalent sets (sets of the same number). It is a short way to add equal numbers. If he has learned that subtraction describes the separation of sets, he can team that division describes the separation of a set into equivalent sets.

Follow the same steps to develop the child's knowledge of these processes as you followed to develop addition and subtraction. The child should:

a) Experience situations that require multiplication and division to solve real problems that affect his life;
b) learn to use the vocabulary and signs necessary to describe these processes orally and in writing; and C) memorize the facts.

PROBLEM SOLVING WITH MULTIPLICATION: You can guide your child's experiences named in steps a and b. Present a problem, allow him to try to solve it, and observe the method used. Do not tell the child the method is wrong. Instead, allow the child to use it, and then show another way.

1. Say; "There will be three people to play with your cars. If we give each per son two cars, how many cars do we need?" After he has tried to find the number, show three sets of two and record this as an addition problem: 2+2+2=6.

   Explain that this can also be written as: 3 x 2 = 6 and read "3 times 2 equals 6." 3 and 2 are called factors and 6 is the product.

   Factor times factor equals product 3x2=6

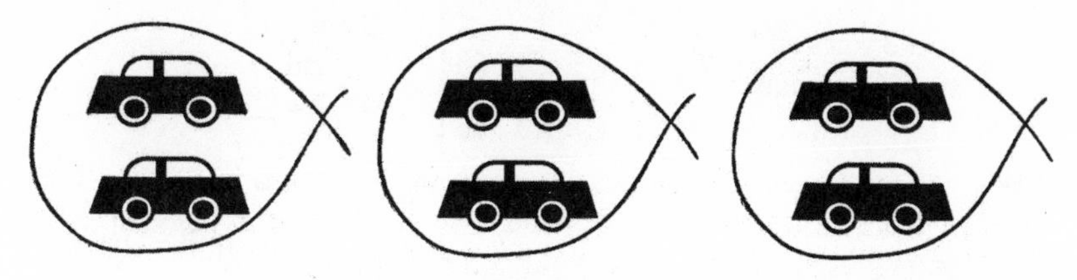

2. Say: There will be two people to color pictures. If we give three pictures to each, how many pictures do we need? Show two sets of three pictures.

Write: 3 + 3 = 6
and 2 x 3 = 6

Read the second sentence as "Two times three equals six." Use the 6 pictures to review: 3 x 2 = 6
and 2 x 3 = 6

3. Show two stacks of four pennies and say: "Here are two stacks of four pen-

nies. Write the multiplication sentence that will tell how many pennies altogether." 2 x 4 = 8

Show four stacks of two pennies and ask for the multiplication sentence.

4x 2 = 8

4. Use an egg canon to show: 2x6= 12
   and: 6 x 2 = 12

Continue to use every situation that you can create to encourage the child to find the product of a number less than six and a number less than ten- through the "five facts." Model each multiplication sentence with objects arranged in rows (⟶)and columns (↓).

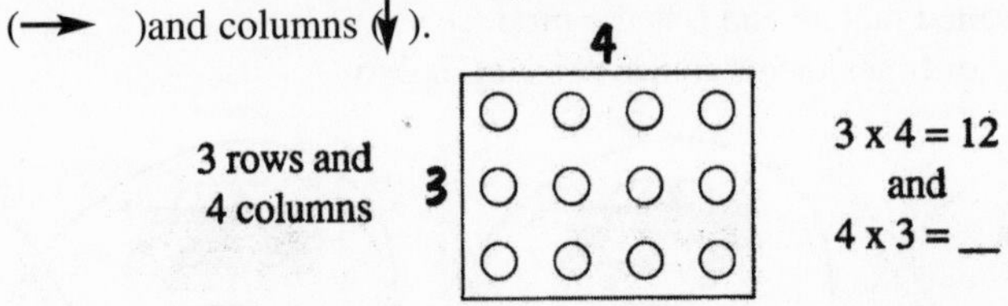

The child will also be learning to write the exercises in vertical form so encourage him to use both forms to record products. Later in the chapter in the section on memorizing facts you will find a list of the facts.

PROBLEM SOLVING WITH DIVISION: Use situations similar to the following to develop the child's understanding of the process of division:

1. Say: "We have six toy cars. There will be two people to play. How many can each have? Let's make two rows with the six cars and see how many will be in each row." Using the cars, lead the child to the conclusion that six divid ed into two equivalent sets will have three in each set. Use an arrangement in rows and columns to help him visualize this order.

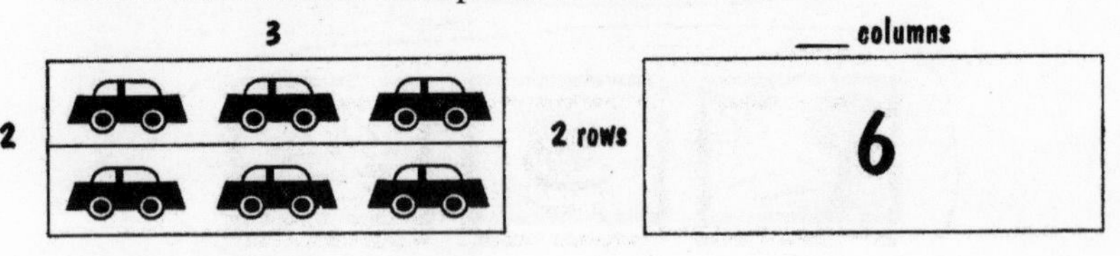

2. Say: "We have twelve crackers. If each person can have four crackers, how many people will get crackers?" Use the arrangement of rows and columns to help the child arrive at the answer.

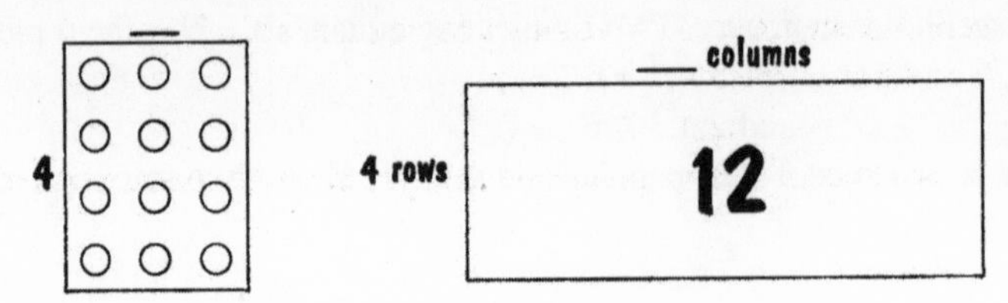

Explain that a short way to express the above problems is to use the division symbol.

As the two exercises above show, there are two aspects of division: sharing and subtraction.

Example of sharing: Show 12 toy cars. Say: "There are 12 toy cars and 4 people. How many cars can each person have?" He may give each person one car at a time. in turn, until all are gone or until there are not enough for all to have another. Show the child that this many be written:

Total cars (12) divided by number of people (4) = (3) cars for each

12 + 4 = 3

Explain that for every division fact there is a matching multiplication fact For this one it is 4 x _ = 12

Example of subtraction: Show 12 toys. Say: "We have 12 toys. Each person gets 4. How many people will get toys? How many sets of 4 can be taken from a set of 12? How many times can 4 be subtracted from 12? The child can give 4 toys to each person until there are no more sets of 4, and then count the people that got toys Show that this may also be written:

Total toys (12) divided by number of people (4) = (3) toys for each. The matching multiplication fact is: _ x 4 = 12.

Continue to create situations that involve the division facts, showing the two aspects of division.

Examples:

1. When planning a shopping trip. say: "I will buy some pencils that cost 9(t each. How many pencils can I buy with 27(?" Give the child 2 dimes and 7 pennies and ask him to show you sets of 9. He will need to trade the 2 dimes for pennies and then place the 27 pennies into sets of 9. Ask him to write a division exercise showing this.

Quotient 3

a.	Dividend		Divisor		Quotient
	27	÷	9	=	3

3 ← Quotient

b. Divisor ⟶ 9)27 ← Dividend

2. Say: "Here is a piece of ribbon that is 24 inches long. How many 8-inch rib bons can we cut from it?" Encourage the child to write the division example that matches the question. 24 + 8 or 8)24
Let the child cut the ribbon into as many 8-inch pieces as possible and check the quotient. Ask the child to show you the matching multiplication sentence: _ x 8 = 24.
3. Show 12 cut flowers and 3 vases. Say: "We have 12 flowers and 3 vases. How many flowers shall we put in each vase?" After he writes the division exercise, suggest that he place the flowers, one at a time. in the vases

4. When planning a snack, say: "We have 21 carrot sticks and 7 people. How many carrot sticks do we have for each person?" After he has written the division exercise with the quotient, suggest that the child place the carrot sticks on seven plates and count the number on each.

Continue to create situations that require the use of the division facts that involve a divisor or a quotient up to 5. Encourage the child to use multiplication facts to match the division facts.

PROBLEM SOLVING WITH MULTIPLICATION AND DIVISION: After the child has used multiplication and division examples, give other problem situations and urge the child to solve each by determining which process to use. Stress that:

a) When equivalent sets are joined, a multiplication sign (x) is used and numbers are multiplied.
b) When equivalent sets are taken away or when a set is divided into a number of equivalent sets, one number is divided by another and ÷ or $\overline{)}$ is used

1. Say: "I have 25 pennies. For how many nickels can I trade these? Each nickel is worth 5 pennies." $5\overline{)25}$ =__
2. Say: "I have 4 nickels. For how many pennies can I trade these?" 4 x5=_
3. Say: "30 people are going on the picnic. 6 can ride in each car. How many cars do we need?" $6\overline{)30}$ =_
4. Say: "There are 7 tables in the restaurant. 4 people can sit at each. How many people can eat in the restaurant at the same time?" 7x4= _.

MEMORIZING THE MULTIPLICATION FACTS: Your child is probably ready to memorize the multiplication facts with products to 45. After understanding comes the need for memorization. Spend a few minutes each day with the child until he has achieved mastery of these facts. Long sessions are self-defeating. Frequent, short sessions of 10 minutes or less are more effective. Continue until the child can give the following products and quotients accurately and rapidly.

0 x 0	1 x 0	2 x 0	3 x 0	4 x 0	5 x 0	6 x 0	7 x 0	8 x 0	9 x 0
0 x 1	1 x 1	2 x 1	3 x 1	4 x 1	5 x 1	6 x 1	7 x 1	8 x 1	9 x 1
0 x 2	1 x 2	2 x 2	3 x 2	4 x 2	5 x 2	6 x 2	7 x 2	8 x 2	9 x 2

0	1	2	3	4	5	6	7	8	9
x 3	x 3	x 3	x 3	x 3	x 3	x 3	x 3	x 3	x 3
0	1	2	3	4	5	6	7	8	9
x 4	x 4	x 4	x 4	x 4	x 4	x 4	x 4	x 4	x 4
0	1	2	3	4	5	6	7	8	9
x 5	x 5	x 5	x 5	x 5	x 5	x 5	x 5	x 5	x 5

For each multiplication fact above, encourage the child to give a matching fact. For example:

6	5
x 5	x 6
30	30

ZERO AS A FACTOR: Of these 60 facts, 15 contain zero as a factor. Most children learn quickly that zero times any number is zero. (0 x 3 = 0,9 x 0 = 0) Thus, he learns 15 facts and has 45 more to learn.

ONE AS A FACTOR: Of these 45 facts, 13 have one as a factor. These, too, arc very easy and when the child leams that one times any number is that number. he has learned 13 facts. (1 x 3 = 3,5 x 1 = 5)

TWO AS A FACTOR: Of the 32 facts without zero or one as a factor, 11 have a factor of 2. These should be rather simple for the child to learn if he has learned the addition facts such as 3 + 3,7 + 7, etc. Use the 0 to 99 square from Chapter 3 to help develop the pattern of these products. Have the child shade the squares with these products. Discuss that all are even numbers. Ask: "How many eyes on 6 people? Hands on 9 people? Feet on 7 people? Ears on 8 elephants?" etc.

FIVE AS A FACTOR: Of the 21 facts without a factor of zero, one, or two, nine contain the factor five. If the child has learned to count by fives, he should have little difficulty with these products. Remind him that each product will have 0 or 5 in the ones place. Ask the child to shade these products on the 0 to 99 square as he shaded the products of two. Ask: "How many fingers are on 6 gloves? 8 nickels are worth how many pennies?"

THREE AND FOUR AS FACTORS: Twelve facts contain none of the factors named above. These contain either three or four and the child will probably need to practice more on these.

3	4	6	7	8	9	3	4	6	7	8	9
x3	x3	x3	x3	x3	x3	x4	x4	x4	x4	x4	x4

After the child has mastered most of the facts, make 3 or 4 index cards of a troublesome fact, with its answer, and position the cards at various places around the house where the child will see them frequently. Occasionally ask for the answer of the multiplication fact.

To provide practice for these multiplication facts, choose games from these listed in Chapter 11.

The Mystery Box, Game 6	Tic Tac The
Fill My Pockets, Game 2	The Trail Game
Pay the Piper, Game 3	Secret Rule
Match up. Game 4	Facts Relay Race, Game 4

Use the child's knowledge of odd and even numbers introduced in Chapter 3 to help him learn a pattern for the multiplication facts.

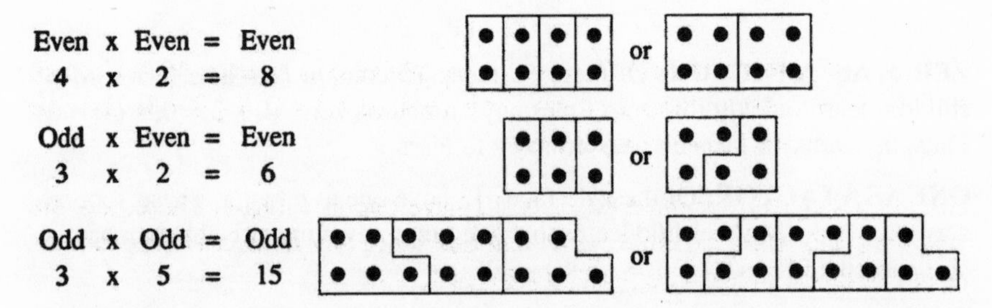

MEMORIZING THE DIVISION FACTS: After your child demonstrates that he knows when to write a division sentence to solve a problem, he is probably ready to memorize the division facts listed below. As with multiplication, keep the work periods short and frequent. Continue practice until he can give the following quotients accurately and rapidly.

$1\overline{)0}$	$1\overline{)1}$	$1\overline{)2}$	$1\overline{)3}$	$1\overline{)4}$	$1\overline{)5}$	$1\overline{)6}$	$1\overline{)7}$	$1\overline{)8}$	$1\overline{)9}$
$2\overline{)0}$	$2\overline{)2}$	$2\overline{)4}$	$2\overline{)6}$	$2\overline{)8}$	$2\overline{)10}$	$2\overline{)12}$	$2\overline{)14}$	$2\overline{)16}$	$2\overline{)18}$
$3\overline{)0}$	$3\overline{)3}$	$3\overline{)6}$	$3\overline{)9}$	$3\overline{)12}$	$3\overline{)15}$	$3\overline{)18}$	$3\overline{)21}$	$3\overline{)24}$	$3\overline{)27}$
$4\overline{)0}$	$4\overline{)4}$	$4\overline{)8}$	$4\overline{)12}$	$4\overline{)16}$	$4\overline{)20}$	$4\overline{)24}$	$4\overline{)28}$	$4\overline{)32}$	$4\overline{)36}$
$5\overline{)0}$	$5\overline{)5}$	$5\overline{)10}$	$5\overline{)15}$	$5\overline{)20}$	$5\overline{)25}$	$5\overline{)30}$	$5\overline{)35}$	$5\overline{)40}$	$5\overline{)45}$

Encourage the child to write a matching fact for each fact above, except those with zero as the dividend. For example: $4\overline{)36}^{\,9}$ and $9\overline{)36}^{\,4}$ are matching facts. As he learns each fact, provide practice with the matching fact. To show that it is impossible to divide by zero, ($0\overline{)3}$) show a set of three objects and ask her to put them into sets with zero in each set. Of course, that is impossible, so dividing by zero is impossible.

FACT FAMILIES: Finding "fact families" may help your child learn multiplication and division facts. Stress that for each pair of factors, there is a family of facts. Sometimes there are two members of the family and sometimes there are four members.

Show examples:

Factors: 3 and 4: $3 \times 4 = 12$ $12 \div 3 = 4$
$4 \times 3 = 12$ $12 \div 4 = 3$

Factors: 5 and 5: $5 \times 5 = 25$ $25 \div 5 = 5$

Name other pairs of factors and encourage the child to write the "fact family" for each pair.

Example: 4 and 6
$4 \times 6 = 24$ $24 \div 4 = 6$
$6 \times 4 = 24$ $24 \div 6 = 4$

3 and 9
$3 \times 9 = 27$ $27 \div 3 = 9$
$9 \times 3 = 27$ $27 \div 9 = 3$

PATTERNS FOR DIVISION: You can help the child learn these 50 facts by asking questions that will lead him to observe simple patterns.

1. Ask: "What is the quotient when you divide a number by one?" (The number itself)
   Examples: $1\overline{)5}$ = 5 $1\overline{)9}$ = 9 $1\overline{)6}$ = 6

   To check the child's understanding of this pattern, show $1\overline{)327}$ and ask him to write the quotient. (327)

   There are ten facts with a divisor of 1. When he has learned this pattern, he will have 40 other facts with divisors 1 to 5 to learn.
2. Ask: "What is the quotient when you divide zero by a number?" (0)
   Examples: $1\overline{)0}$ = 0 $2\overline{)0}$ = 0 $3\overline{)0}$ = 0 $4\overline{)0}$ = 0

   To check the child's understanding of this pattern, write $283\overline{)0}$ and ask him to write the quotient. (0)

   When he has learned the facts with 1 as a divisor and 0 as a dividend, there will be 36 more of the facts to learn. He will also have learned 4 other facts: $6\overline{)0}$ = 0 $7\overline{)0}$ = 0 $8\overline{)0}$ = 0 $9\overline{)0}$ = 0

   Stress: If the dividend is zero the quotient is zero.
3. Ask: "What is the quotient if we divide a number by itself?" (1)
   Examples: $2\overline{)2}$ = 1 $3\overline{)3}$ = 1 $4\overline{)4}$ = 1

   To check the child's understanding of this pattern, show $283\overline{)283}$ and ask him to write the quotient. (1)

   Of the 36 facts not described in examples 1 and 2 above, there are 4 in which the divisor and dividend are the same and therefore 1 is the quotient.

Learning this pattern will reduce the number of facts for the child to learn to 32. It will also help him learn 4 other facts that are not listed.

$6\overline{)6}$ $7\overline{)7}$ $8\overline{)8}$ $9\overline{)9}$

4. With the same dividend the divisor and quotient may be reversed, except when the quotient is zero. Do not divide by zero.

Examples: $2\overline{)6}$ with quotient 3 so $3\overline{)6}$ with quotient 2 $3\overline{)12}$ with quotient 4 so $4\overline{)12}$ with quotient 3

Of the 32 facts not explained in examples 1, 2, and 3 above, six have a matching fact with a divisor less than 6. When the child has learned these 12 facts, there will be 20 more with divisors to five for the child to learn. Learning to identify matching facts (reversing the divisor and quotient) will help him learn 16 facts with divisors greater than five.

Examples: $3\overline{)27}$ with quotient 9 so $9\overline{)27}$ with quotient 3 $4\overline{)32}$ with quotient 8 so $8\overline{)32}$ with quotient 4

DRILL AND PRACTICE: If your child has learned the multiplication facts, he should have little difficulty learning the division facts. Keep the drill and practice of these facts as interesting as possible by choosing games from the list below that are described in Chapter 11.

Tic Tac Toe
Facts Relay Race, Game 5
Pay the Piper, Game 4
Match up. Game 5
The Trail Game

Carefully note the quotients that the child gives slowly or incorrectly. Repeat the games often, using only the facts that give difficulty. Show pleasure at any effort and success.

Telling Time

A MINUTE: When your child becomes aware of the clock face and the movement of the hands on non-digital clocks, explain that it takes one minute for the long hand to move from one little mark to the next. Suggest that he count the number of minutes between each pair of numerals. (5)

To give the child a better idea of the length of a minute, encourage him to watch the second hand move around the dial and explain that length of time as one minute. Ask the child to:

a) See how much of a picture he can color in one minute. You do the timing;
b) write as many sums, differences, products, or quotients as he can in one minute;
c) try to straighten his room (put his toys away) in one minute; and

d) close his eyes and try to guess the length of one minute. Tell the child to open his eyes when he thinks one minute has passed.

FIVE MINDTES: Show the time on the hour (9:00 o'clock) on the demonstration clock and have the child name the time. Move the minute hand to 1 and give the time as "five minutes past (nine)." Continue to move the hand to 2, 3. 4, ...II and talk about the times: ten minutes past nine; fifteen past nine, etc. Show the child that he can begin at twelve on the clock and count by fives to the minute hand to find the minutes past the hour.

After he is able to read the times from the clock, show how these times are recorded.

3:05 Five minutes past three.
4:30 Thirty minutes past four.
5:40 Forty minutes past five.

He can now read these times from a digital clock.

During your daily activities, ask the child often to tell you the time. Also ask such questions as:

1. "It is now 2:10. Your TV program begins in 20 minutes. What time does it begin?" Encourage the child to begin at the minute hand and count the min utes by fives to 20 to determine where the minute hand will be in 20 minutes and thus to (ell the time.
2. "It is now 7:45. In 45 minutes it will be time for you to go to bed. What time are you going to bed?" Use the demonstration clock and discuss the fact (hat as the minute hand passes 12 another hour is named.
3. "We are going to eat lunch in about 20 minutes. What time is it now? What time will we eat lunch?"

HALF HOUR: Sometimes your child will hear the time named as "Half past" the hour. To introduce this term, use circles showing the fractional pans.

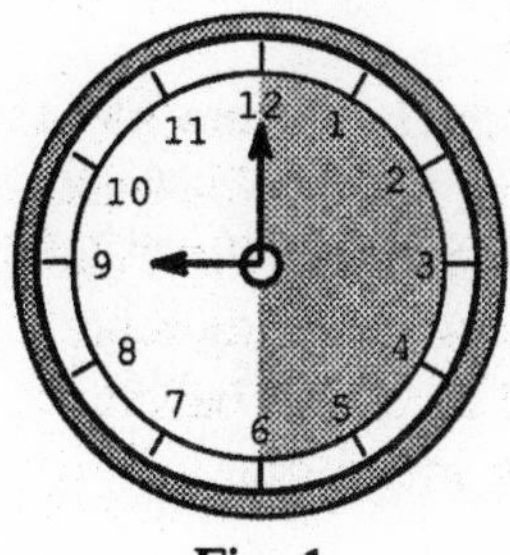

Fig. 1

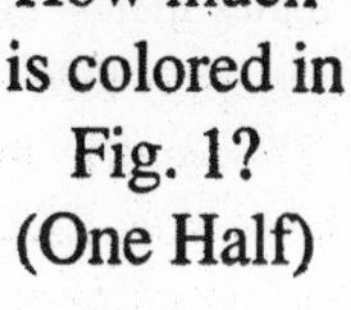

How much
is colored in
Fig. 1?
(One Half)

Fig. 2

Explain that the minute hand in Fig. 2 has moved half way around the clock face, from 12 to 6, and he can give the time as: thirty past nine; nine thirty; half past nine.

Liquid Measures

CUPS, PINTS, QUARTS, GALLONS: In your child's presence, refer to milk cartons and other containers by their measures. Use activities similar to the following to help the child become acquainted with these names.

1. Say: "Please put this pint jar on the shelf."
2. Say: "Please bring a quart jar of applesauce from the shelves."

If necessary, show a quart jar and say, "This is a quart jar. I'd like you to bring me another quart jar." Be sure that there are jars of the other sizes on the shelves and that the child must make a choice.

Ask such questions as: "How many cups will a pint hold? How can we find out?" Suggest that the child use water to see how many cups it takes to fill one pint, and complete: 1 pint holds the same amount as _ cups, so 1 pint equals _ cups.

Follow the same process to find these equivalent measures:

1 quart holds the same amount as _ pints, so 1 quart = __ pints. 1 gallon holds the same amount as _ quarts, so I gallon = __ quarts. 1 quart holds the same amount as _ cups, so 1 quart = _ cups.

Encourage the child to use multiplication to complete exercises similar to the following:

1. 3 pints = __ cups. (If necessary, explain that since 1 pint holds exactly the same as 2 cups, 3 pints = 3 times 2 cups; 3 x 2 = 6)
2. 4 quarts = _ pints (1 quart = 2 pints, so 4 quarts = 4 x 2 pints = 8 pints.)
3. 5 quarts = __ cups (1 quart = 4 cups, so 5 quarts =5x4 cups = 20 cups.)
4. 3 gallons = _ quarts (1 gallon = 4 quarts, so 3 gallons =3x4 quarts = 12 quarts.)

Provide as much help as needed to check each answer by filling the larger container with the smaller container and counting the number required.

LITERS: Explain to your child that a liter is also a liquid measure like cups, pints, quarts, and gallons. If you have a liter container, fill it with water, and have the child pour as much as possible into a quart bottle. Ask: "Which is more, a liter or a quart?" (a liter) This metric measure is becoming more common, so check your containers. Some soft drinks and alcoholic beverages are now sold in liter bottles. Use these containers for the measuring activity and stress that a liter is a little more than a quart.

Addition and Subtraction Two-Digit Numbers with Regrouping

Learning to add and subtract two-digit numbers with regrouping (carrying and borrowing) probably will not be difficult for your child if he has learned:
1. The addition and subtraction facts with sums to 18; and
2. the place value of digits in our number system.

ADDITION: When planning a shopping trip, encourage your child to find how much money you, or he, will need to buy two items that are each priced between 10¢ and 99¢. After the child has attempted to answer the question, model each problem with dimes and pennies and stress that ten pennies will be exchanged for one dime.

Examples:

1. Say: "I want to buy a box of paper clips for 69¢ and a large envelope for 27¢. How much money do I need?" If he has difficulty finding the sum, help him model the problem with dimes and pennies. When he has counted the pen nies, encourage the child to exchange ten pennies for one dime and record as follows:

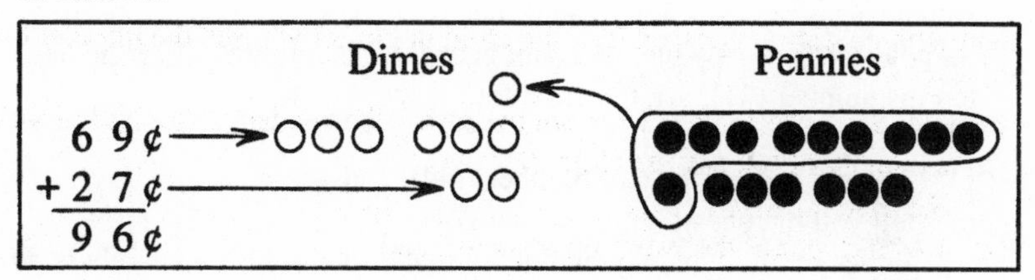

2. Showing two toys in an ad, say: "How much money will you need to buy this toy for 76¢ and this toy for 84¢. If he has difficulty, help the child model the problem as in example 1: trade ten pennies for a dime and ten dimes for a dollar, and record as: 76¢ + 84¢ = 160¢, equals $1.60

If the child continues to need models to solve these problems, use other materi-als that are easily grouped by tens: popsicle sticks, tongue depressors, nails.

Example: Show 25 nails (2 bundles of ten and 5 nails). Say: "Here are 25 nails. If we put 7 nails with them, how many will there be? Write the addition problem and show me what happens with the nails."

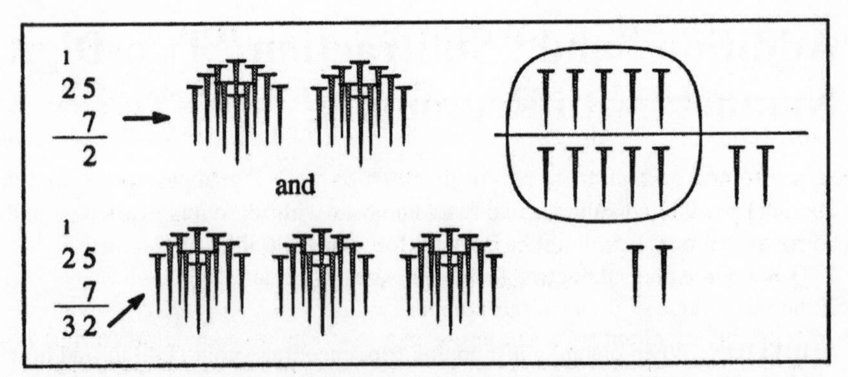

Continue to create situations that present real problems for the child to solve by adding these numbers. Use the models if the child is confused.

JUST FOR FUN: Most children like a little mystery and can be motivated by an opportunity to look for clues and solve a puzzle. Here are two addition exercises that will probably challenge your child to look for patterns and, therefore, increase his interest in the computation.

1. This example presents a simple pattern and should increase the child's con fidence for the search involved in example 2.
   Ask the child to write a two-digit number with both digits less than five and with no digits repeated. Tell the child not to let you see the number that he writes. Sample: 31 is acceptable. 33 is not because 3 is repeated. After he has writ ten his number (31), ask the child to:
   a) Reverse the digits (13);
   b) find the sum of the two numbers (44); and
   c) tell you the ones digit of his sum (4).

   Explain to him that you will tell him the sum of his two numbers. When he tells you that the ones digit is 4, you know that the sum is 44 because both digits must be the same. If he disagrees, you will know that he has failed to follow directions or has made an error in addition.

   Reverse roles. You write a two-digit number with digits less than five, reverse the digits, and find the sum. Tell the child the ones digit and ask him to guess the sum. If he can guess your sum, the child probably recognizes the pattern and can play the same "trick" on other members of the family or friends. If he can't identify the sum, repeat the process.
2. This activity is slightly more difficult.

   Ask the child to write a two-digit number with both digits greater than four. (58) Then ask the child to:
   a) Reverse the digits (85);
   b) find the sum of the two numbers (143); and
   C) tell you the ones digit of the sum (3).

Tell the child you will name the sum of her two numbers. (143) When he tells you that the ones digit is 3, you know that the sum is 143 because the hundreds digit will always be 1 and the tens digit will be the sum of the ones and hundreds digits. The child says 3 is the ones digit. You know that the hundreds digit is 1 and that 3 + 1 = the tens digit (4) So the sum must

be 1 4 3
+

Another example:

Child's number 69 Ones digit in answer is 5; Hundreds digit
Reversed: 96 is 1; 5 + 1=6, so the sum is 165

Reverse roles as in example 1 to allow the child to practice giving the directions and determining the answer. He can then "play the trick" on family members or friends.

SUBTRACTION: Your child should be successful with learning the process of subtracting with regrouping, if:

a) He knows the subtraction facts with sums to 18; and
b) he is aware that the number to be "taken away" is the bottom number of the problem. Model problems similar to the following:

1. Place 3 dimes and 2 pennies on the table. Tell the child that he can have 7¢ if he writes (he numbers and sign that describe what happens when 7 is taken away from 32. 32¢
-7¢

   Ask the child to model the problem by removing the pennies. Ask such questions as: "Can you take 7 pennies from 2 pennies? (No) What should you do so that you can remove 7 pennies? (Exchange one diire for ten pen nies.) Now, how many pennies do you have? (12) How many dimes? (2) Show this. (Encourage the child to write the regrouping above the top num ber.) Can you remove 7 pennies, now?" (Yes) Help record this if necessary.
2. Place 4 rings of ten and 5 individual snap beads on the table and ask the child to give you 29 beads to make a necklace. Suggest that he write the exercise that will tell how many will be left.

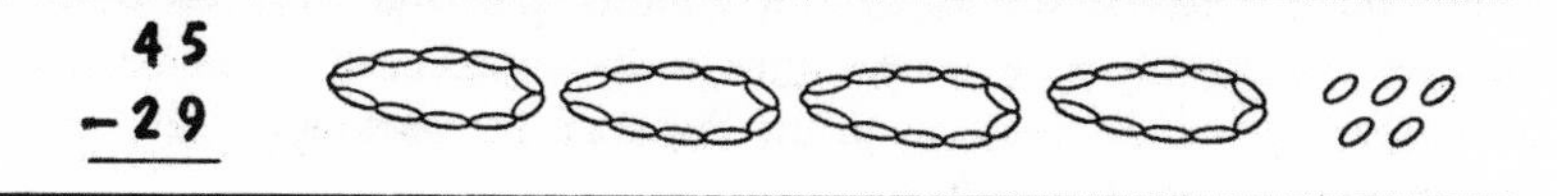

Ask such questions as: "Can you take 9 beads from 5 beads? (No.) Can you group the beads so that you can remove 9? (Change a ring of 10 to ten ones.) Show this with the beads. Show it with numbers."

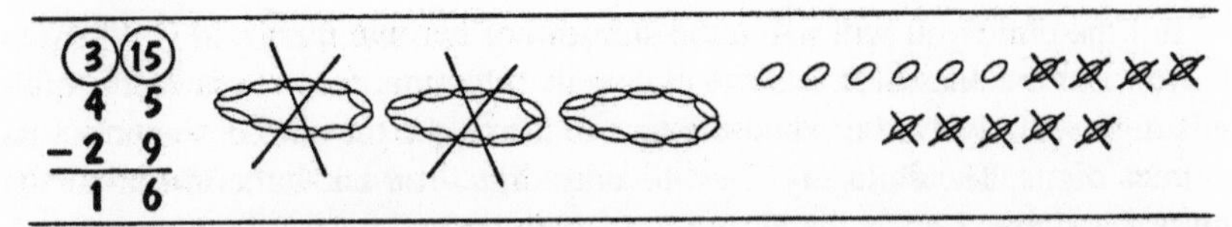

Create other situations that will require the child to change a group of ten to ten ones before he removes the specified number. Urge the child to perform the action and record the results as a subtraction exercise. It is very important for the child to meet situations that require addition and subtraction with regrouping to find answers that occur in daily life. Use every opportunity to present these situations to the child.

JUST FOR FUN: Use this activity to motivate your child to solve more subtraction problems with two digits and to help him become more aware of patterns in our number system.

Tell the child to write a two-digit number with different digits. (Correct: 38 or 97. Incorrect: 55 or 99.) Tell the child not to let you see the number that he writes.

After he has written the number (83), tell the child to:

a) Reverse the digits and write another number (38);

b) subtract the smaller number from the larger, and

c) tell you the ones digit of the difference (5). When he tells you the ones digit, mentally subtract that number from nine to find the tens digit of Wher answer. (9 - 5 = 4, so the difference is 45.)

Examples:

	Child's Secret Number	Reversed:	Problem:	
a)	16	61	61 -16 -5	 (9-5=4) Answer. 45
b)	95	59	95 -59 -6	 (9-6=3) Answer 36

If the child thinks that he knows how you found the difference, reverse roles and let the child guess the difference of the numbers that you write. If he has determined the pattern, encourage the child to "play the trick" on other family members and friends. If he shows signs of discouragement before he detects the pattern, give the child clues, such as: "Subtract the ones digit from nine. Does this help you find the answer?"

SOLVING PROBLEMS WITH ONE AND TWO STEPS: After your child has learned a computation process, the most difficult part of solving a problem using that process is determining what to do. Now that the child has learned to add and subtract two-digit numbers, use every opportunity to encourage the child to solve problems using these processes.

Examples:

1. When shopping, ask questions similar to these: "Here is a sticker for 12< and one for 9(. How can we find how much they cost together? (Add) How much money do you have? (Be sure he has less than $1.00.) How can you find how much you will have left if you buy the two stickers? (Subtract the amount they cost together from the amount I have.) Find out how much they cost and how much you will have left if you buy the stickers."
2. Now find a similar situation and ask the child one question that will require
t two steps to find the answer. "You have S12. How much will you have left
if you buy this toy for $3 and this toy for $5? How can you find out?" (Add the cost of the two toys and subtract the sum from $ 12.)
3. "That bike costs S78. You have S25 in your savings account and $14 in your little bank. How much more do you need to buy the bike? How can you find out?"

Continue to look for problem solving situations that arise in your child's life that require addition and subtraction of 2-digit numbers. Stress the question: How do we find the answer?

Play Store, Game 3 from Chapter 2.

Summary

The exercises in Chapter 4 have introduced the place value of 3-digit numbers to your child. This background has allowed the child to work with amounts of money from $1.00 to $9.99 and to add and subtract 3-digit numbers. Regrouping in addition and subtraction has been limited to 2-digit numbers to allow for time to develop complete understanding.

Skills with fractions and telling time have been expanded and activities with liquid measures have been introduced.

The meaning of multiplication and division has been stressed in this chapter and techniques to aid the child in memorizing the facts with 0 to 5 as factors have been presented. Use as many games and enjoyable activities as possible to promote mastery of these facts.

5

USING NUMBERS TO 9,999

The activities in this chapter are designed to reinforce additional math skills for your child and to extend these skills to include the numbers to 9,999. These skills are usually introduced in third-grade texts.

Grade Three.

1. Counting and writing numerals to 9,999;
2. addition and subtraction of three-digit numbers with regrouping;
3. fractions: equivalent to one; equivalent fractions; comparing unit frac tions;
4. multiplication and division facts to 9 x 9 and 819;
5. multiplication and division of two- and three-digit numbers by one-digit numbers;
6. measurement of weights: pounds, ounces, grams, kilograms;
7. measurement of length: feet, yards, meters;
8. measurement of perimeter of squares and other rectangles;
9. measurement of area of rectangles; and
10. use of thermometers.

Take the above list of skills with you to teacher conferences. Ask the teacher to check those skills for which you should be providing experiences for your child.

Numbers to 9,999

Your child has probably learned that in any number each place is ten times the value of the place on its immediate right.

Thousands	Hundreds	Tens	Ones	
			3	= 3 ones
		3	0	= 30 ones or 3 tens or 10 x 3
	3	0	0	= 300 ones or 30 tens or 3 hundreds or 100 x 3

Write 1,000 and ask the child if he knows a name for 10 hundred. If he does not know, show it on the chart and tell the child the name. (One thousand) Ask: "One thousand names how many hundreds? (Ten) How many tens? (One hundred) How many ones? (One thousand)

Write other numbers from 1,000 to 9,999 and have the child read each number. Give numbers orally (four thousand, six hundred seventy-two) and ask the child to write the numerals that name each number. Stress the comma between thousands and hundreds.

Play Number Relay, Games 1 and 2 from Chapter 11.

Addition and Subtraction Three-Digit Numbers with Regrouping

Your child should be ready to add and subtract three-digit numbers with regrouping (carrying and borrowing) if he has learned:

1. The addition and subtraction facts with sums to 18;
2. to add and subtract two-digit numbers with regrouping; and
3. the place value of three- and four-digit numbers.

ADDITION: Ask questions of interest to the child that will motivate him to add these numbers. First, give a problem that requires regrouping only from ones to tens.

"We drove our blue car 257 miles last week and we drove our white car 236 miles. How many miles did we drive both cars?"

If he has difficulty, continue giving problems that involve only this one regrouping until he has this process mastered.

Next give problems that require regrouping from tens to hundreds.

"One of your new toys costs $2.63 and the other costs $5.52. What is the total cost of the toys?"

If necessary, use dollars, dimes and pennies to model the problem and don't go to the next step until he can work this type of problem easily.

After he can work problems like the two above, ask questions that will require the child to regroup from ones to tens and from tens to hundreds.

"If you earn $4.75 on Monday and $3.25 on Wednesday, how much will you earn in the two days?"

Reinforce what he has learned by providing as many interesting real-life problems as possible that require the process.

Money provides an excellent model for these problems (exchange 10 pennies for a dime and exchange 10 dimes for a dollar) and is also a subject of interest to most children.

SUBTRACTION: As with addition, ask questions of interest to your child that will motivate him to subtract these numbers. Check to see if he has mastered the skill. First ask a question that will require only regrouping from tens to ones.

"The toy you want costs $3.45 and you have $1.28. How much more money do you need to buy the toy?"

Be sure he can solve problems that involve only this one regrouping before going to the next step.

Next give a problem that requires regrouping from hundreds to tens.

"Your friend has 246 marbles (or any other small items) and you have 152. Who has more? How many more?"

After the child has learned to do the single regroupings, ask questions that require him to regroup both from hundreds to tens and from tens to ones.

"Sarah jumped rope 365 times and Kirn jumped rope 298 times. Who jumped more times? How many more?"

Continue to ask questions that will require your child to subtract with regrouping. Money also provides an excellent model for these problems. (Exchange one) dime for ten pennies and one dollar for ten dimes.)

Play Store. Game 4 from Chapter 11.

CHECKING SUBTRACTION: Your child may have learned to check the accuracy of the answer 10 a subtraction problem. He does this by adding the difference to the subtrahend. If the sum is equal to the minuend, the difference is correct.

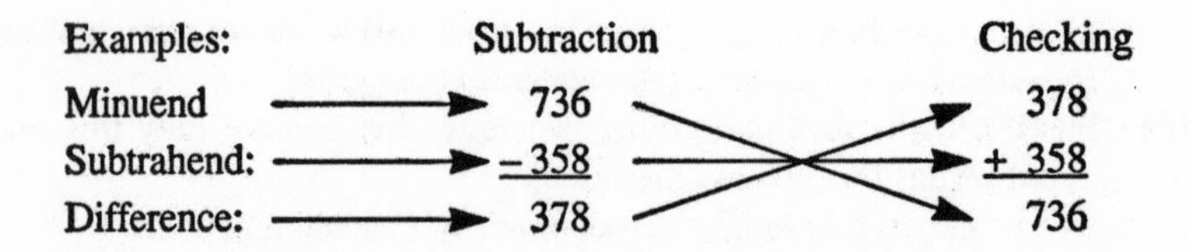

Encourage the child to check all of his subtraction problems for accuracy.

JUST FOR FUN: Ask your child to:

- **a.** Write a secret three-digit number with each digit different from the others (Correct: 528 or 397. Incorrect: 355 or 633 or 343);
- **b.** reverse the digits (528 becomes 825);
- **c.** subtract the smaller number from the larger $\begin{array}{r} 825 \\ -528 \\ \hline 297 \end{array}$ and;
- **d.** name the digit that is in the ones place of the difference. (7) After he names this digit, tell the child that you can now tell the answer even though you do not know the numbers of her problem.

Your solution:

The tens digit of the difference will always be 9 and the sum of the hundreds digit and the ones digit will always be 9 (2 + 7 = 9). To determine the difference, subtract the ones digit named by the child from 9 to find the hundreds digit and place 9 in the tens place.

If the child says this is not the answer, then you know that he has not followed directions or has made an error in subtraction.

Other examples:

$$\begin{array}{r} 613 \\ -316 \\ \hline 297 \end{array} \qquad \begin{array}{r} 723 \\ -327 \\ \hline 396 \end{array} \qquad \begin{array}{r} 841 \\ -148 \\ \hline 693 \end{array} \qquad \begin{array}{r} 793 \\ -397 \\ \hline 396 \end{array}$$

PRACTICE: Continue to present the child with situations that require addition and subtraction with and without regrouping. Make the problems as related to daily life as possible. For each, encourage the child to determine the process, addition or subtraction. Assist by asking whether sets will be joined (numbers added), or separated or compared (numbers subtracted).

Refer to the chapters in Part II for other real-life situations that require addition and subtraction.

Fractions Equal to One; Equivalent; Comparing

1	Numerator	2
2	Denominator	3

Ask your child to write a fraction with 3 as the numerator and 4 as the denominator 3/4. Show the child the parts labeled above if he has difficulty.

Write the following: 1/2, 2/3, 3/4. Ask the child to name the fraction with a numerator of 2 (2/3); a numerator of 3 (3/4}; a numerator of 1 (1/2); a denominator of 2 (1/2); a denominator of 4 (3/4)); a denominator of 3 (2/3).

Continue to refer to the terms of the fractions as the numerator and the denominator.

Example:

> Cut an orange into four equal parts. Tell your child that he can have X. Ask: "What does the numerator tell? (The number of parts he can have.) What does the denominator tell?" (The number of pans altogether.) Ask the same questions about the fraction that tells how much of the orange is left. (3/4)

EQUAL TO ONE: When following a recipe calling for one cup of an ingredient. ask your child how many 1/2- cup measures it will lake to make one cup. Encourage the child to fill the 1 - cup measure with water from a 1/2 - cup measure and help record the results as follows: 1/1 = 1

Follow the same procedure with one third- and one fourth- cup measures and record the results. 3/3 = 1 and 4/4 = 1

Show a pie or pizza cut into 7 pieces of the same size. Ask your child to give the fractional name of one piece ('/^); two pieces (¥1); three pieces (-^); etc. Ask for the fractional name of the whole pie as shown by the seven pieces. ('/;)

Use the egg carton and the fractional parts of the carton prepared as described in Chapter 4 to find fractional names for one. Ask the child to fit the blue parts in to the whole carton. Ask: "How many parts equal one whole carton?" (Two.) Each part is what fraction of the carton?" ('/^) Suggest that he write the fractional name for one in halves. (1/2)

Follow the same procedure for thirds (red canon) and fourths (yellow carton).

Encourage the child to color one egg carton orange and cut it into six equivalent parts; color another brown and cut it into 12 equivalent parts. Use the parts of the carton to complete: $\frac{\_\_\_}{6}=1$ $\frac{\_\_\_}{12}=1$

Ask: "What can you tell me about the numerator and the denominator of a fraction that is equal to one?" (They are equal.)

Give these fractions for the child to complete without the use of concrete objects.

$$1 = \frac{\;}{3} \qquad 1 = \frac{\;}{5} \qquad 1 = \frac{\;}{7} \qquad 1 = \frac{\;}{8}$$

EQUIVALENT FRACTIONS: Use a recipe that calls for ½ cup of an ingredient and explain to the child that you only have a clean ¼-cup measure. Ask if it is possible to use that to measure ½ cup. Allow the child to pour water from the ¼-cup measure into the unwashed ½-cup measure to find how many fourths are equal to one half. Suggest that he record this by completing this sentence:

$$\frac{\bigcirc}{4} = \frac{1}{2}$$

Show a pie or pizza cut into four equivalent parts. Ask the child for the fractional name for one part (¼); two parts (2/4); three parts (¾);; four parts (4/4). Ask the child to cut each fourth into two parts, count all of the parts (8), and give the fractional name of one part. (⅛) Ask him to complete the following sentence:

$$\frac{1}{4} = \frac{\bigcirc}{8}$$

Fold a sheet of paper into three equivalent parts. Tell the child to color ⅓.

Fold the sheet to show sixths.

Ask how many thirds are colored.

(1) How many sixths are colored?

(2) Complete: 1/3 = ___/6

2/6 = ___/3

Use the same activity to show:

$$\frac{1}{2} = \frac{\bigcirc}{4} \qquad \frac{1}{2} = \frac{\bigcirc}{8}$$

Place the blue parts of an egg carton, as prepared in Chapter 4, into the whole carton and have the child name one part. (One half) Ask the child to find how many fourths fit into one half and complete: 2/4 = ½

Follow the same process to have these sentences completed:

$$\frac{1}{4} = \frac{\bigcirc}{12} \qquad \frac{2}{6} = \frac{\bigcirc}{3} \qquad \frac{2}{3} = \frac{\bigcirc}{12}$$

COMPARING FRACTIONS: To show the importance of being able to determine which fraction is larger, ask the child which he would like you to give him, one fifth of a dollar or one tenth of a dollar. Place ten dimes on the table and ask him to arrange them into five equal sets. Discuss that each set is one fifth of a dollar. Ask how much is in each set. (20¢) Record: ⅕ of 100 = 20¢. Ask the child to place the dimes into ten sets and call each set one tenth of a dollar. Ask how much is in one tenth of a dollar. (10¢) Record:1/10 of 100 = lO¢. Did he choose the larger fractional part?

Before sharing an apple, orange, cookie, or other treat, ask: "Which would you rather have, one half or one third? Why?" Show one half is larger by cutting one treat into halves and another of the same size into thirds and comparing the fractional pans. When comparing fractional parts of two items, be sure the items are equivalent in size. Write these sentences and ask the child to complete them by writing '/^ and !6 in the proper places. Use the divided treats if necessary.

____ is greater than ____.

____ is less than ____.

Show1/2-, 1/3-, and a 1/4 -cup measures to the child and ask him to order them from least to greatest-left to right. Encourage the child to write the names in the same order. (1/4 1/3 1/2)

Write these fractions in random order 1/3, 1/2, 1/12, 1/4, 1/6. Ask the child to rewrite them in order from greatest to least, left to right. Allow the use of egg canons for comparison if necessary. (1/2, 1/3, 1/4, 1/6, 1/12)

After he has put the fractions in order, ask if he notices any order to the denominators. Lead him to the observation that they are in order from least to greatest. Ask the child to tell about the numerators. After he observes that all of the numerators are one (1), ask if he knows a name for fractions with a numerator of one. These are called unit fractions.

Ask where 1/5 would fit in the sequence of fractions that he ordered. (The fraction 1/5 would fit between l/4 and 1/6.With activities similar to those above, lead the child to the conclusion that as the denominators of unit fractions get larger the size of the fraction gets smaller.

Your child has probably been using the mathematical symbols for "is greater than" and "is less than." These symbols were introduced in Chapter 3. Review the fact that the larger end of the symbol is on the side of the larger number.

Example: 15 > 13 "Fifteen is greater than thirteen." 13 < 15 "Thirteen is less than fifteen."

These symbols are also used to compare fractions. Write:

a. $\frac{1}{2} \bigcirc \frac{1}{4}$ b. $\frac{1}{4} \bigcirc \frac{1}{2}$ c. $\frac{1}{10} \bigcirc \frac{1}{12}$ d. $\frac{1}{4} \bigcirc \frac{1}{3}$

Guide the child as he places the right symbol in each.

Have each sentence read:

a. One half is *greater than* one fourth.
b One founh is *less than* one half.
c. One tenth is *greater than* one twelfth.
d. One fourth is *less than* one third.

Use egg cartons to check examples a, b, and d. Fold 2 sheets of paper (one in 10 equal pans and one in 12 equal pans) to check example c.

Multiplication and Division Through the "Nine Facts"

MULTIPLICATION: If your child has memorized these facts through the "five facts", he has very few more to learn. The shaded products in the table below were presented in Chapter 4. The new products are unshaded. It may appear that there are 15 more products for the child to learn but, remember, if the factors are reversed, the product remains the same.

For example: 7 x 6 = 42 and 6 x 7 = 42

For 7x6= _: In the table below find 7 in the left column and 6 in the top row. Follow the two arrows to the intersecting square to find the product. (42)

For 6x7= _: Find 6 in the left column of the table and 7 in the top row. Follow the arrows to the intersecting square to find the product. (42)

So: 7x6=42 and 6x7=42

Practice:

7x8=_ 9x6=_ 8x9=_ 8x7=_ 6x9=_ 9x8=_

X	0	1	2	3	4	5	6	7	8	9
0	0	0	0	0	0	0	0	0	0	0
1	0	1	2	3	4	5	6	7	8	9
2	0	2	4	6	8	10	12	14	16	18
3	0	3	6	9	12	15	18	21	24	27
4	0	4	8	12	16	20	24	28	32	36
5	0	5	10	15	20	25	30	35	40	45
6	0	6	12	18	24	30	36	42	48	54
7	0	7	14	21	28	35	42	49	56	63
8	0	8	16	24	32	40	48	56	64	72
9	0	9	18	27	36	45	54	63	72	81

Since six products are repeated and four are not repeated (6 x 6 , 7 x 7, 8 x 8, 9 x 9), there are only ten new facts for the child to learn.

6	6	6	6	7	7	7	8	8	9
x6	x7	x8	x9	x7	x8	x9	x8	x9	x9

Create situations that require multiplication similar to those created for learning facts through the "five facts" in Chapter 4. Encourage the child to model any fact if he is unsure of the product.

Extend any of the games listed in Chapter 4 to include the six to nine facts. Play Back to Back, Game 2, and Buzz from Chapter 11.

You can simplify the nine facts by explaining to the child that the digits in the answer to each of these always add up to 9; i.e. 9 x 9 = 81 (8 + 1 = 9); 9 x 3 = 27 (2 + 7 = 9); 9 x 5 = 45 (4 + 5 = 9); etc. Nine times any number results in a tens number that is one less than the other number. The sum of the tens and ones will equal nine.

9 x 3 = 27 (2 is one less than 3) 9 x 6 = 54 (5 is one less than 6)
9 x 4 = 36 (3 is one less than 4) 9 x 7 = 63 (6 is one less than 7)

DIVISION: If your child has learned the division facts through the "five facts," he has very few more to learn. The following are the new facts that he needs to learn.

$$6\overline{)42} \quad 6\overline{)48} \quad 6\overline{)54} \quad 7\overline{)56} \quad 7\overline{)63} \quad 8\overline{)72}$$

$$6\overline{)36} \quad 7\overline{)42} \quad 8\overline{)48} \quad 9\overline{)54} \quad 7\overline{)49} \quad 8\overline{)56} \quad 9\overline{)63} \quad 8\overline{)64} \quad 9\overline{)72} \quad 9\overline{)81}$$

Extend the games listed in Chapter 4 to include the facts listed above. Play Back to Back, Game 2, from Chapter 11. Watch closely as the child plays; note the quotients that he gives incorrectly or hesitantly. Encourage the child to repeat his favorite games and use the facts that are difficult. Show pleasure when he masters a new fact.

FACT FAMILIES: Show and discuss the fact families for multiplication and division. Stress that for each pair of factors, there is a family of facts. Sometimes there are two members in the family and sometimes there are four members.

Examples:

Factors:			
	6 and 7:	6 x 7 = 42	42 ÷ 6 = 7
		7 x 6 = 42	42 ÷ 7 = 6
	8 and 9:	8 x 9 = 72	72 ÷ 8 = 9
		9 x 8 = 72	72 ÷ 9 = 8
	7 and 7:	7 x 7 = 49	49 ÷ 7 = 7

Encourage the child to write the members of the fact family for each pair of factors from 6 to 9. Finding these fact families may help the child memorize the multiplication and division facts.

Multiplication: Two- and Three-Digit Numbers by One-Digit Numbers

Multiplying two- and three- digit numbers by one-digit numbers will be much easier for the child if he has learned the multiplication facts through 9x9. You can model this process with dimes and pennies or other sets of tens and ones.

MULTIPLYING TENS: To simplify the process of multiplication, begin by creating situations that require the child to multiply tens.

1. Say: "If I give you 10? each day for three days, how much will I give you?" Suggest that he record the process that tells how much money he will get.

$$3 \times 10 = 30 \quad \text{or} \quad \begin{array}{r} 10 \\ \underline{\times\ 3} \\ 30 \end{array}$$

2. Say: "How much will you get if I give you 20¢ a day for four days?" Ask the child to record the process and encourage the child to tell you how he thinks through the process. He may have difficulty doing this, so don't expect too much. If there is trouble, suggest a method for a few problems. Show 20¢ as 2 dimes or 2 sets of 10 pennies. Ask: "How much is 3 times 2 dimes?" (6 dimes or 60¢)
3. When planning for shopping, name an item that costs 40< and say: "How much will 3 of these cost?" Use dimes to model if necessary.

MULTIPLYING TENS AND ONES WITHOUT REGROUPING: Extend the process of multiplication to include tens and ones by creating situations and questions similar to the following. At this point, be sure that the total number of ones is less than ten. Have no regrouping (carrying).

1. Say: "I want 3 pieces of yam. Each piece must be 23 inches long. How much yam do I need? Write the exercise." If he has not learned the process, allow her to model the problem with sticks.

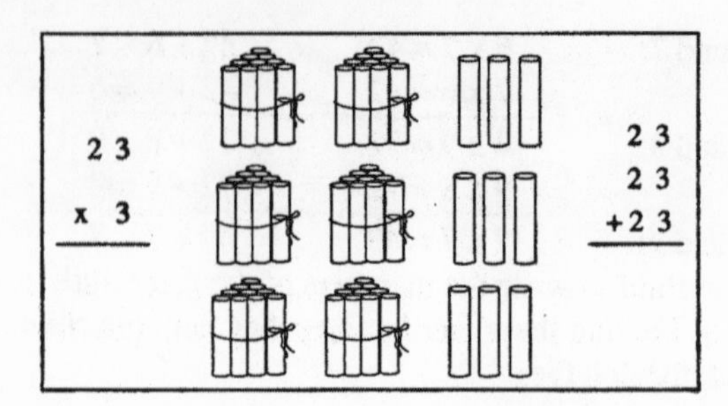

Allow the child to use the long form first if he has difficulty, and encourage him to use the short form when he is able to see the relationship between the two forms of the problem.

Long Form:		Short Form:	
Tens	Ones	Tens	Ones
2	3	2	3
x	3	x	3
	9	6	9
+6	0		
6	9		

2. When preparing a garden say: "We have 4 rows for tomatoes and we can put 12 plants in each row. How many plants do we need?"

$$\begin{array}{r} 12 \\ \times\ 4 \\ \hline 8 \\ +40 \\ \hline 48 \end{array} \quad \text{or} \quad \begin{array}{r} 12 \\ \times\ 4 \\ \hline 48 \end{array}$$

Continue to create situations that require this process.

MULTIPLYING TENS AND ONES WITH REGROUPING: Present problems similar to the following to provide practice for multiplication with regrouping. Use models to show the grouping of ten ones and carrying if the child has difficulty with the process.

1. Show the price of an item in the grocery ads and ask the child to find how much a number of them will cost. Say: "How much is one can of frozen apple juice? (48¢) How much will two cans cost?"

Long Form: or Short Form:

$$\begin{array}{rl} 48 & \\ \times\ 2 & \\ \hline 16 & (2 \times 8) \\ +80 & (2 \times 40) \\ \hline 96 & \end{array} \quad \text{or} \quad \begin{array}{r} {}^{1}48 \\ \times\ 2 \\ \hline 96 \end{array}$$

To model this problem, show 2 sets of 48 (4 dimes and 8 pennies in each set). Have the child count the pennies and exchange as many sets of ten pennies for dimes as possible. Record as shown in the long form above. Have the child count all of the dimes and record the number as shown in the short form above.

2. Say: "I need 7 macrame cords and each cord must be 36 inches long. How much cord do I need?"

Long Form: or Short Form:

$$\begin{array}{rl} 36 & \\ \times\ 7 & \\ \hline 42 & (7 \times 6) \\ +210 & (7 \times 30) \\ \hline 252 & \end{array} \quad \text{or} \quad \begin{array}{r} {}^{4}36 \\ \times\ 7 \\ \hline 252 \end{array}$$

3. Have (he child count your pulse for 1 minute. Ask: "How many times did my heart beat in 3 minutes? 5 minutes?"
4. Tell the child the number of miles your car will travel on one gallon of gaso line. "How many miles will it travel on 8 gallons (or any number less than 10)?"

MULTIPLYING HUNDREDS, TENS, AND ONES WITH REGROUPING: Create problem situations that will require your child to multiply 3-digit numbers with regrouping. Use bundles of sticks or hundred-squares, ten-strips, and unit-squares to model the problems if necessary. Examples:

1. Say: "If we travel 250 miles each day for 3 days on our vacation, how many miles will we travel?"
2. Pour three to nine cups of beans into a container. Suggest that family mem bers guess the number of beans. After the guesses are recorded, give the child the cup and ask how he can use it to find the approximate number of beans. If necessary, suggest that he count the number of beans in one cup, find (he number of cups of beans altogether, and multiply.
3. Ask the child if he knows the number of days in one year. (365) Ask: "If you are 8 years old, how can you find how many days old you are? (Multiply.) Find your age in days."
4. Say: "If you sleep 9 hours each night for one year, how many hours will you sleep in a year?" (9 x 365)

Division: Two- and Three-Digit Numbers by One-Digit Numbers

Your child needs to know how to subtract and multiply in order to learn to divide two- and three-digit numbers by one-digit numbers. He also needs some speed and accuracy with multiplication and division facts.

DIVIDING TENS AND ONES: For meaningful practice of division, take advantageof the child's experiences with handling money. A combination of dimes and pennies makes a good model for a division problem.

1. Place 2 dimes and 3 pennies on the table. Ask the child to write the amount of money. Suggest that he share the money equally between 4 people-him self and three others. Ask the child to find (he amount each will get. Ask:

Allow the child to use the long form first if he has difficulty, and encourage him to use the short form when he is able to see the relationship between the two forms of the problem.

```
Long Form:          Short Form:
Tens  Ones          Tens  Ones
 2     3             2     3
x      3            x      3
-----------         -----------
       9             6     9
+6     0
-----------
 6     9
```

2. When preparing a garden say: "We have 4 rows for tomatoes and we can put 12 plants in each row. How many plants do we need?"

```
  12           12
x  4    or   x  4
----         ----
   8           48
+ 40
----
  48
```

Continue to create situations that require this process.

Multiplying Tens and Ones with Regrouping: Present problems similar to the following to provide practice for multiplication with regrouping. Use models to show the grouping of ten ones and carrying if the child has difficulty with the process.

1. Show the price of an item in the grocery ads and ask the child to find how much a number of them will cost. Say: "How much is one can of frozen apple juice? (48¢) How much will two cans cost?"

```
Long Form:                 Short Form:
                             1
  48                         48
x  2               or      x  2
----                       ----
  16  (2 x 8)                96
+ 80  (2 x 40)
----
  96
```

To model this problem, show 2 sets of 48 (4 dimes and 8 pennies in each set). Have the child count the pennies and exchange as many sets of ten pennies for dimes as possible. Record as shown in the long form above. Have the child count all of the dimes and record the number as shown in the short form above.

2. Say: "I need 7 macrame cords and each cord must be 36 inches long. How much cord do I need?"

```
Long Form:                 Short Form:
                             4
   36                        36
x   7              or      x  7
-----                      ----
   42  (7 x 6)              252
+ 210  (7 x 30)
-----
  252
```

digit numbers offers many challenges. The process involves two other operations (multiplication and subtraction) and many steps. The fact that we perform the operation from left to right is often confusing because the other operations are usually performed from right to left. Children sometimes have trouble determining the location of the digits in the quotient.

Examples:

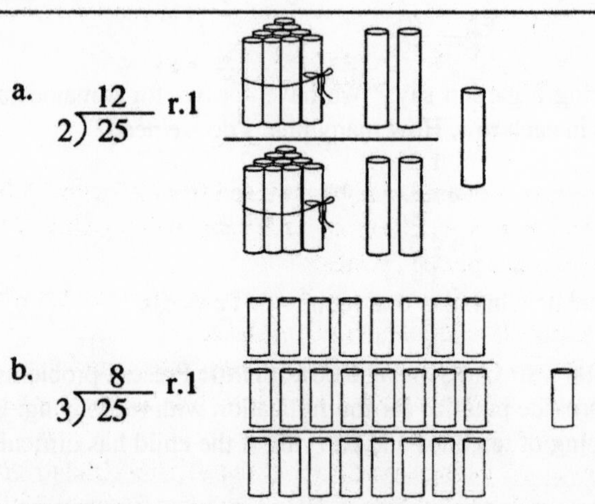

Use models to show that:

1. In example a, 2 tens can be divided into 2 equivalent sets without being regrouped to ones. The result is 1 ten in each set, so the 1 is written in the tens place.
2. In example b, the divisor is greater than the number of tens, so the tens must be regrouped with the ones. The result is 8 ones in each set, so the 8 is writ ten in the ones place.

Use models of hundreds, tens and ones to explain division of three-digit numbers by one-digit numbers. The models should clarify the steps of the process and the place value of the digits of the quotient. Use hundred-squares, ten-strips, and unit-squares (described in Chapter 4) to model problems like the following:

```
   5 8 R.2
4)2 3 4
  2 0
    3 4
    3 2
      2
```

1. Say: "We are dividing 234 into 4 partss. Can we put a hundred in each pan? (No) so we divide 23 tens by 4." This will show why the 5 goes in the tens place. Continue questioning for the rest of the problem.
2. Guide the child with questions: "How many hundreds for each of the two parts?" (1) This will show why the 1 is placed 2 in the hundreds place.

```
  1 1 7
2)2 3 4
  2
    3
    2
    1 4
    1 4
```

Provide as many practical experiences for your child as possible that require division to solve problems that occur in daily life.

Example:

1. Say: "The book has 245 pages. If you are going to read it in 7 days, how many pages must you read a day?"
2. "1 have $6.23 to divide between 3 children. How much for each child?" Model with dollars, dimes, pennies.
3. "It is 982 miles to our vacation spot We must drive it in 4 days. How far must we drive each day?"
4. "There are 78 pennies in the bank. For how many nickels can we exchange these pennies?"

Measurement of Weight

POUNDS AND OUNCES: Your child is probably familiar with the pound as a measure of weight. Interest in his own growth and development makes the child keenly aware of how many pounds he weighs. Show Lb. as a short way to write "pound."

Show one pound of coffee and tell the child how many of these it would take to be as heavy as he is. Ask the child to lift items (a book, box of cereal, toy, bag of sugar) alternately with the coffee and tell which items he thinks weigh as much as one pound, the weight of the coffee. Have the child weigh the items on the bathroom scale to check the guesses.

Show the child a 5-lb. bag of flour and ask him to tell how many bags of flour it will take to weigh as much as he does. Encourage the child to divide his weight by 5.

Explain that weight is also measured in ounces. Show oz. as a short way of writing "ounces." Encourage the child to find items in the kitchen with the weight given in ounces. Suggest that he use a food scale to find the weight of other small items in ounces: a carrot, a few ounces of meat, any small amount of food on waxed paper. Ask if the scale shows 1 -Lb. and if he can find how many ounces are in one pound. Ask: "How many ounces does the pound of coffee weigh? How many ounces are in five pounds of sugar?" Encourage the child to multiply 16 by 5.

GRAMS: Many items in your kitchen have their weights given in grams. Suggest that your child find some of these items and check their weights on the food scale. After checking many of these weights, ask the child to guess the number of grams that other items weigh and then weigh the items to check the guesses.

Measurement of Length

FEET AND YARDS: Show a 12-inch ruler and tell your child that this is a measure called afoot. Ask: "How many inches are in one foot?" (12) Ask the child to use this ruler to find how many people in the family have a foot that is a foot long. Encourage him to use the foot ruler to compare the length to a foot and record with is greater than (>), is less than (<), and is equal to (=) signs.

a) My foot _ 1 foot.
b) Mother's foot _ 1 foot
c) Daddy's foot _ 1 foot
d) John's foot __ 1 foot.

Suggest that he use an 18- or 36- inch ruler to measure the length of the family members' feet and record the lengths as inches, or feet and inches.

Examples: Daddy's foot = 14 inches = 1 foot 2 inches
Mother's foot = 12 inches =1 foot

Mark the child's height on a wall and have the child use the foot ruler to find his height in feet and inches or just inches.

Example: My height = _ inches = _ feet _ inches.

Ask the child to find the height of other family members and record these heights in the same manner.

Urge (he child to measure other items in feet and inches.

Examples:

1. The length and width of a table;
2. the length of a jump rope; and
3. the length of a board to be used in construction.

After the experiences described above, show a yard stick and ask if he knows the name of the measure. If not, tell the child it is one yard. Ask for the number of inches in one yard; the number of feet.

Ask for many lengths to be measured and ask your child which measuring tool will be better to use, the foot ruler or the yard stick.

Examples:

1. The length of a hallway (yardstick);
2. the width of a TV set (foot ruler);
3. the distance between horseshoe pegs (yardstick); and
4. the length or width of your yard (yardstick).

Suggest that the child find which family members have an arm that is one yard long. Use the signs >, <, and = to compare the lengths to 1 yard.

Record: My arm is _ 1 yard. Mother's arm _1 yard. Daddy's arm _ 1 yard.

Ask: "Do you think Daddy is a square?" Show the child how to measure daddy's height and arm span to determine if they are the same. thus forming a square. Ask the same question about other members of the family and also about the child.

After your child has had many measuring experiences with inches, feet. and yards, play a game of "Guestimation." Each player writes a secret estimate of a distance named by the leader. The estimates are given to the leader and the players measure the distance. The player whose estimate is closest to the actual measure wins a point or a prize.

METERS: If your child has learned to measure distances with centimeters, he should easily learn to measure with meters. Show a meter stick and ask how many centimeters make a meter. (100) This can be related to 100 cents making a dollar. To practice measuring with meters, use activities similar to those described for measuring with yards.

Perimeter

When the child has drawn, painted, or colored a picture on a square of paper, tell your youngster it is "pretty enough to be framed." Suggest that he find how many inches of ribbon it will take to frame the square. Notice how he finds the distance around the square. Does he measure each side and add the lengths? Does he measure one side and multiply that length by 4? Ask if he knows one word that means the distance around a figure. (Perimeter) If he doesn't know, tell the child.

As he finds the perimeter of squares, ask if the child can find the perimeter if he only knows the measure of one side. (Multiply that measure by four.)

Stress that a square is one kind of rectangle. Show other rectangular pictures (not squares) and ask the child to find the amount of ribbon needed to frame each. Ask if he can find the distance around a rectangular window if he only measures two sides. (Add the length and width and multiply (he sum by two.)

Ask the child to find the perimeter of:

1. Your yard in feet. to determine how much fence will be needed to enclose it;
2. room in feet or yards, to determine how much baseboard is needed to go around the room;
3. a handkerchief to determine how many inches of lace will be needed to trim it.

4. a cushion to determine the number of inches of fringe needed.to go around it.

Use other opportunities to have your child find the perimeter of squares and rectangles.

Area

When your child begins to measure area, he must learn to count square units. If any floor, wall, or cabinet top in your home is covered with square tiles of equal size. use a cord to outline a rectangular area that is 3 tiles by 4 tiles and ask the child to name the area in square tiles (12) and then use the cord to show other rectangles with an area of 12 square tiles. (1 by 12,2 by 6)

Suggest that the child cut a rectangle from graph paper with an area of 12 square units. Compare this area to the area of square tiles and stress the importance of specifying the size of the units in order to tell the actual size of an area.

Cut one square inch from paper. Have the child measure the length of each side and identify it as I inch. Discuss: If each side of the square is 1 inch. the area is expressed as 1 square inch.

Show a rectangular piece of paper that is 2 inches wide and 4 inches long. Ask the child to measure its sides in inches and tell the area. Encourage the child to mark each inch in the perimeter and draw square inches to check the area.

Continue with other rectangles with length and width given in inches until the child identifies that length times width equals the area of a rectangle. (1 x w = A)

Cut one square foot from brown wrapping paper. Ask the child to measure each side of the square and name the area in square feet (1 square foot)

Use cord to outline an area two feet wide and three feet long. Ask the child to give the area in square feet. (2x3=6)

Measure an area 3 feet by 3 feet and outline it with cord. Ask the child to give the length and width in feet and to give the area in square feet (9) Ask for the length and width in yards (1) and the area in square yards. (1 square yard)

Urge your child to assist you as you find the area of a room to determine the amount of carpet to buy, or the area of a wall to prepare for buying wall paper. He may not have the computation skills (multiplying by a 2-digit multiplier), but can assist with measuring and tell you how to find the area (multiply the length by the width). Involve the child in many of these activities.

When your child is studying metric units of measurement, use these activities with square centimeters and square meters.

Thermometers

Provide your child with as many opportunities as possible to read a thermometer-Fahrenheit or Celsius. Explain the number of degrees represented by each mark and encourage the child to:

1. Read and record the temperature when the thermometer is in the shade; and
2. move the thermometer to a sunny spot, wait 5 or 10 minutes, and read and record the temperature. Ask: "Did the mercury (liquid) move up or down as it got wanner? (Up.)

Using the Fahrenheit thermometer, ask the temperature at which water freezes. (32°) If the child doesn't know, place a thermometer in water in the freezer and encourage the child to check it constantly until the water begins to freeze and to name the temperature at this point. Discuss the temperature at which water boils (212°) and help the child use a candy thermometer to check this.

Show Fahrenheit temperature, by pictures or on a thermometer, below zero and ask how these are written. If necessary, show that they are written with a minus sign. (-3° is read: "Three below zero.")

Use the same procedures to show that water freezes at 0° and boils at 100° on the Celsius thermometer.

Show pictures of outdoor activities in books or magazines and ask your child to choose between two temperatures that might exist as the activity is taking place.

Examples:

A person skiing: 25°F. or 98°F.
Children playing baseball: 75°F.or-5°F.

SOLVING PROBLEMS WITH ONE AND TWO STEPS: After learning the skills presented in this chapter, your child should be able to solve problems that require one and two steps. The most difficult part will be determining which process or processes to use to find the answer. Begin by asking for the answer to the first step and then for the answer to the second step. After the child has had experience with answering one step at a time, ask one question that will require the use of two steps to find the answer.

Examples:

1. "It is 520 miles to our vacation spot. If we drive 225 miles each of the first two days, how far will we need to drive the third day? How can we find the distance we will drive the first two days? (Multiply) How can we find how far we will have to drive the third day? (Subtract) Find the number of miles we will drive on the third day."
2. "The bike costs $89 and you have $35. If you earn $6 each week mowing

grass, how many weeks will you have to work to have enough money to buy the bike? How can you find how much more money you need? (Subtract) How can you find how many weeks you need to work? (Divide) Find out."
3. "You had $3.47 this morning. You spent $1.49 and earned $2.50 today. How can you find how much you have now?" (Subtract and add)

As you look for meaningful situations that require the use of the computation skills your child has learned, always stress the question: How will you find the answer?

Summary

Your child's background has been expanded to include numbers to 9,999. This has allowed him to add. subtract, multiply and divide with these greater numbers. As the child's computation skills have increased, his measurement skills have also increased.

Continue to provide games and activities that motivate your child to memorize multiplication and division facts with factors to 9.

6
EXTENDING THE USE OF NUMBERS

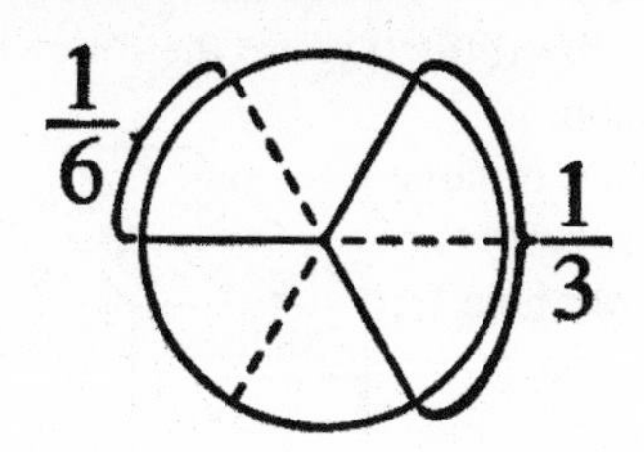

The activities in this chapter will help develop and reinforce many skills that your child will learn in grade three and above. Many are extensions of those introduced in Chapters 4 and 5. Refer to those chapters if your child has difficulty and needs more developmental work in any particular skill.

This chapter includes the following skills and concepts:

1. Reading and writing whole numbers through billions;
2. reading and writing decimal numbers through hundredths;
3. adding and subtracting any set of whole numbers;
4. multiplying and dividing 4-, 5-, and 6-digit whole numbers by 1-digit whole numbers;
5. adding and subtracting decimal numbers in tenths or in hundredths;
6. multiplying and dividing 1- and 2-place decimals by 1-digit numbers;
7. adding and subtracting fractions with like denominators;
8. multiplying proper fractions;
9. using combinations of operations to solve two-step problems including find ing averages;
10. finding the perimeter of a polygon;
11. finding the area of rectangles and other shapes;
12. finding the volume of rectangular-shaped boxes;
13. expressing measures in miles and kilometers; and
14. making and reading scale drawings.

Before reaching the intermediate grades, your child will probably have mastered the basic facts in addition, subtraction, multiplication and division of whole numbers. Mastery means being able to give an immediate response. If he continues to have difficulty memorizing the facts, play Match Up, Back to Back, Buzz, and Dealer Calls the Answer from Chapter 11 and continue other activities suggested in previous chapters.

Place Value

WHOLE NUMBERS: A good grasp of our place value system will increase your child's understanding of numbers greater than 9,999. Ask him to read 7 to 12 digit numbers to you. If he has difficulty, use the following activities to help him see the patterns that develop.

Draw a chart on a file folder or large piece of paper and separate the periods with a heavy dark line, as shown here.

hundreds	tens	ones	hundreds	tens	ones	hundreds	tens	ones
Millions			Thousands			Units		

Point out that each period, or set of three places, includes ones, tens, and hundreds. This pattern continues for each succeeding period.

To read any number with more than three digits, begin on the left, say the number formed by the digits to the left of each comma, say the name of the period at the comma, and then read the number of units.

For example:

The digits are separated frorn right to left into sets of three. Therefore, the first period on the left will have one, two, or three digits:

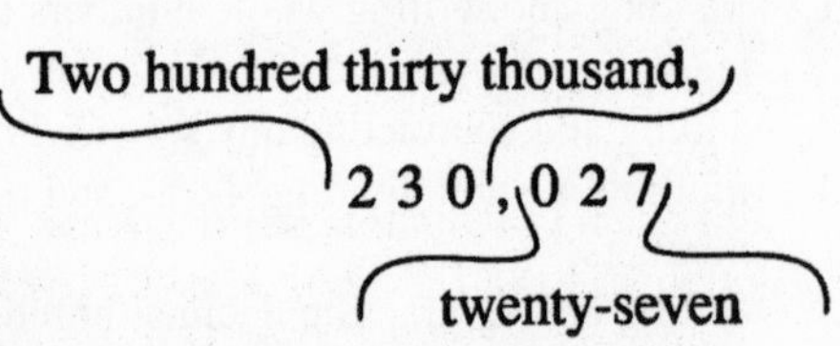

1,436.782 *31*,762 236,741

Stress that and is not used in the number name unless there is a decimal point or fraction.

Example:

342,106 Say: "Three hundred forty-two thousand, one hundred six."

Wrong: Three hundred and forty-two thousand, one hundred and six.

25.016 Say: 'Twenty-five thousand, sixteen."

Discuss topics of interest to your child that involve large numbers. For example:

1. The seating capacity of the stadium used by the child's favorite sports team.
2. The mileage on the odometer of your car.

3. The distance from the sun to the earth.
4. The population of the largest city in your state or a large city near your home.
5. Readings on water and electric meters in your home.

Have your child find large numbers in the newspaper or encyclopedia and read them aloud. To correct any errors, have him write the numbers on a place value chart and read the number in each period from left to right

To help the child comprehend the size of a million, discuss the following topics:

1. If a person earned $30 every day from the day that person was bom, he or she would have to live more than 91 years to earn one million dollars.
2. One million minutes is almost 2 years (more than 1.9 years).
3. A child's heart beats almost one million times a week. It takes about 19 years for it to beat one billion times.

Write a 6-digit number (635,437) on the chart. Point to a digit (5) and name the value (5 thousands). Point to another digit (6) and ask the child to name the place value (6 hundred thousands). Repeat for other digits.

Write other 6- to 9-digit numbers on the chart in order to make it easier for the child to see the groups of three and the name of each period. Ask him to read each number and give the place value of the digits as you name them. Stress that large numbers are easy to read if we look at the digits in groups of three and know the names of the periods.

JUST FOR FUN: Ask your child to research the names of the remaining periods on this place value chart and then to write them on the chart. (World Book: Decimal Numeral System)

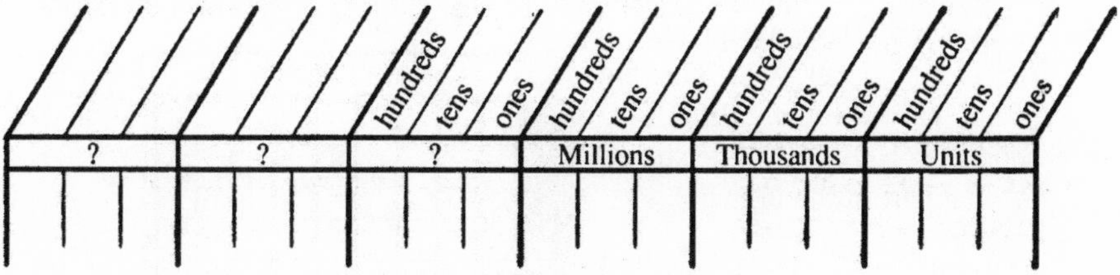

						hundreds	tens	ones	hundreds	tens	ones	hundreds	tens	ones	hundreds	tens	ones
?			?			?			Millions			Thousands			Units		

Play Building the Greatest Number from Chapter 11.

DECIMALS: Young children's experiences with decimal numbers may currently deal mostly with money, but as the use of metric measures becomes more common, decimals will be presented earlier in the curriculum and encountered more often in their everyday experiences. Your child, with the understanding that in our place value system each place in a number is 10 times the value of the place to the immediate right, should have little difficulty with decimal numbers.

You have material at your fingertips to show the meaning of these decimal numbers, material that has meaning for your child: money, metric measures, a car odometer. You can use these materials to help him develop the understanding and to practice the use of these numbers.

Examples:

1. Show a one dollar bill, some dimes and some pennies. Ask: "How many dimes make a dollar? Each dime is what part of a dollar? (1/10 or .10) How many pennies make a dollar? Each penny is what part of a dollar?" (1/100 or .01)
2. Show 2 dollars, 5 dimes, and 6 pennies and ask the child to write the amount. ($2.56) Ask: "There arc how many whole dollars? What does the decimal point do? (Separates the whole dollars from parts of a dollar.) There arc how many tenths of a dollar? (5) How many hundrcths?" (Accept 6 or 56.)
3. Show 3 dollars and 2 dimes. Ask the child to write the amount. ($3.20) Then ask questions similar to those above. If the child does not recognize .20 as 20 hundrcdths, trade the dimes for 20 pennies and ask: "What part of a dol lar is each penny? (One hundredth) What part of a dollar is 20 pennies? (20 hundrcdths) Does .2 equal .20? (Yes. just as 20 pennies equals 2 dimes.) Does a zero on the right change the value?" (No.)
4. To stress the use of zeros in decimal numbers, show a place value chart (see right) and ask the child to record amounts of money on the chart. Discuss each amount. $2.34: "How many whole dollars? (2) How many tenths of a dollar? (3) How many hundredths?" (4) $2.07: "How many whole dollars? (2) How many tenths of a dollar? (None) How many hundrcdths?" (7) Show the following amounts and ask similar questions: $0.24, $0.30, $0.03, $0.40, $0.04

Tens | Ones | Tenths | Hundredths

5. Cut several 10 by 10 squares from graph paper. Ask: "One big square has how many little squares? (100) Each little square is what part of the big square? (.01 - one hundredth) How many columns of ten are in each big square? (10) Each column is what part of the big square?" (. 1 - one tenth)

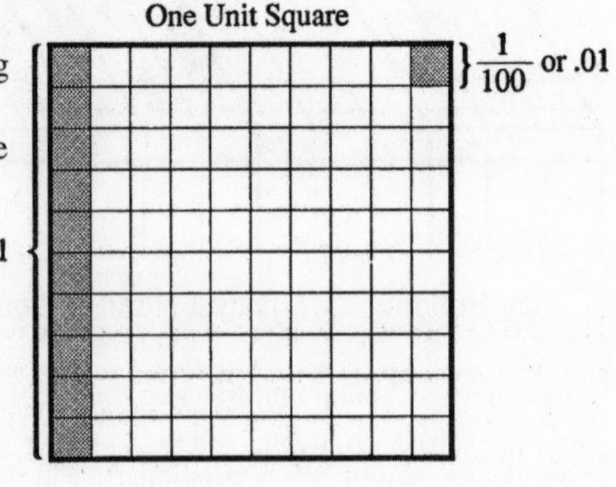

To show the relationship of dollars and cents to any decimal number, erase the dollar sign from a money number ($3.20) and ask the child to model that number by coloring squares (3.20-3 whole squares and 2 columns of the fourth square). Continue with other money numbers-$0.27, $1.04, $0.50, $0.05.

6. Show a meter stick and ask: How many centimeters are in one meter? (100) Each centimeter is what pan of a meter? (One hundredth.) Write the num ber." (.01 or 0.01.)
7. Cut ten 10 cm lengths of drinking straws (pipe cleaners or other firm mate rials) and lay them end-to-end beside a meter stick. Ask: "Are all straws the same length? How long is each? (10 cm) How many does it take to make a meter? (10) Each straw is what pan of a meter? (One tenth) Write that num ber. (.1 or 0.1) Another name for 10 cm is decimeter. One decimeter is one-tenth of a meter just as a dime is one- tenth of a dollar."
8. Tell the height of a family member in centimeters and ask the child to change it to meters. (125 cm = 1.25 m) Give other distances (length or width of his room, height of a table or bike) in centimeters and ask for the measure in meters.
9. Name lengths similar to those in exercise 8 in decimeters and ask the child to write each length in meters. 13 dm = _ m (1.3) 7 dm = _ m (.7)
10. Ask your child to read the mileage (kilometers) from the odometer of your car and to record it so that you can check the miles per gallon (kilometers per liter). Did he record tenths correctly?
11. Explain that a decade is ten years and say: "One year is what pan of a decade? Write that number. (.1) Change 27 years to decades." (2.7) Have your child change other numbers of years to decades and decades to years. **Examples**: 3.4 decades = __ years (34) 6 years = _ decades (.6)
12. Explain that a century is 100 years and say: "One year is what pan of a cen tury? Write that number. (.01) 25 years is what pan of a century? (.25) Change 208 years to centuries." (2.08)

Computation

ADDITION AND SUBTRACTION OF WHOLE NUMBERS: After mastering the basic addition and subtraction facts and learning to add and subtract with regrouping, your child should be able to use these operations with large numbers as welL Watch for these trouble spots in addition.

1. He forgets to add a regrouped (carried) number. Suggestion: Encourage the child to write each regrouped number rather than keep it in his head.
2. He regroups the wrong number. For example, he writes the tens place num ber in the ones place and carries the ones place number to the tens.

Suggestion: Use a place value chart for the problem and sum, in order to call attention to the ones place, tens place, hundreds place, etc.

	TH	H	T	O
		1	1	
	4	3	6	8
	7	4	3	2
+	6	1	2	4
1	7	9	2	4

3. He makes errors in addition problems that are written in horizontal form. Suggestion: Have the child rewrite the problem in vertical fonn, paying particular attention to the place value of each digit

79+380+4519

	TH	H	T	O
			7	9
		3	8	0
+	4	5	1	9

Watch for these trouble spots in subtraction.

1. Your child subtracts the smaller number from the larger, regardless of whether it is on the top or the bottom. Suggestion: Stress: Subtract the bottom number from the top number. If the bottom number is greater, regroup (borrow). Encourage the child to show every regrouping.

7	10	17	2	12
~~8~~	~~1~~	~~7~~	~~3~~	~~2~~
−5	6	8	1	7
2	4	9	1	5

2. He makes errors when zeros are in the minuend (top number).

Suggestion: Have him:

(8 0) 4 − 3 6   (6 0 0) 3 − 2 5 4 6   (6 0) 1 3 − 2 5 3 1

a) Circle the zero, or zeros, and the first non-zero digit on the left

7 9 14 / (8 0) ~~4~~ / − 3 6   (80 tens − 1 ten = 79 tens)

5 9 9 13 / (6 0 0) ~~3~~ / − 2 5 4 6   (600 tens − 1 ten = 599 tens)

b) regroup 1 from the circled number to the number from whichhe is subtracting; and

5 9 11 / (6 0) 1 3 / − 2 5 3 1   (60 hundreds − 1 hundred = 59 hundreds)

c) subtract

7 9 14 / (8 0) ~~4~~ / − 3 6 / 7 6 8   5 9 9 13 / (6 0 0) ~~3~~ / − 2 5 4 6 / 3 4 5 7   5 9 11 / (6 0) 1 3 / − 2 5 3 1 / 3 4 8 2

3. **He regroups when it is not necessary. Suggestion: Encourage the child to make all regroupings before he begins to subtract. In the following example it is not necessary to regroup to the ones, but it is necessary to the tens. Encourage your child to check subtraction often by adding the answer to the subtrahend (bottom number).**

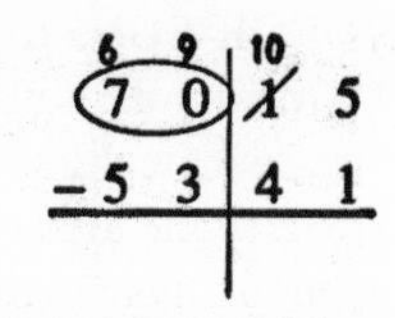

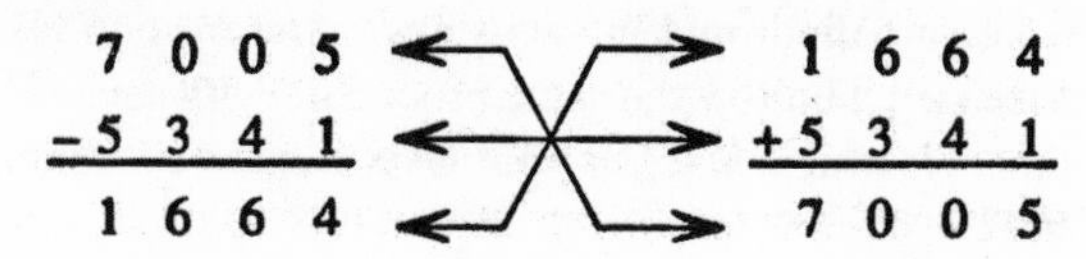

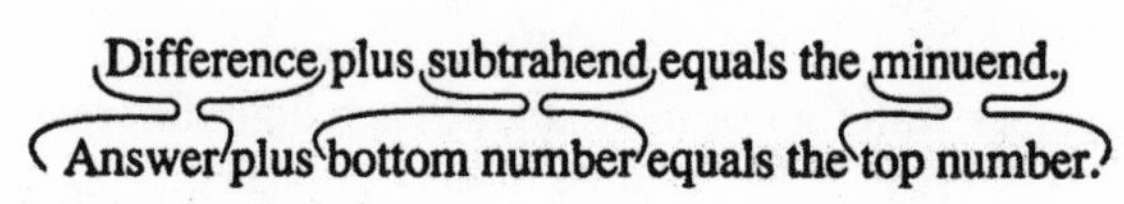

Activities similar to the following will provide opportunities to add and subtract in meaningful real-life situations.

1. Ask your child to record the mileage on the odometer of your car when you begin and finish a trip and to find how many miles you traveled.
2. Tell your child the whole number odometer reading and the distance to your destination. Ask him to find what the odometer reading will be at the end of the trip.
3. Tell the present reading and the previous reading from your electric or natu ral gas bill. Ask the child to find the number of units (kilowatt hours or hun dred cubic feet) your family used during the month. Ask him to keep this record for several months to find in which months you used the most (least).
4. Encourage your child to locate facts in encyclopedias, textbooks, or newspa pers, and find the answers to questions that require addition or subtraction.
   a) Which is farther from the earth, the sun or the moon? How much far ther?
   b) Which weighs more, an elephant or a whale? How much more?
   c) How many people does our football stadium seat? How many empty seats were there last weekend?
5. Use a mileage chart to locate distance to different cities. Ask questions sim- lar to the following:
   a) How far is it from our city to Detroit?
   b) Which is farther away, Detroit or Miami? How much farther?
   c) If we drive to Detroit and then to Minneapolis, how far will we drive?
   d) Today we drove 325 miles on our trip to Detroit. How much farther will we have to drive?

JUST FOR FUN: To practice addition, have your child:

1. Write a 3-digit number,	724	1151
2. reverse the digits and add;	+ 427	+ 1511
3. reverse the digits again and add; and	1151	2662

4. continue this process until (he sum is a palindrome.

A palindrome is a number (or word) that reads the same in cither direction.

MOM, DAD, RADAR, MADAM, and POP are palindromes. The example shown above makes a palindrome in two steps. Now try 546. This will give a palindrome in three steps. Follow the same procedure with several other numbers and count the steps. How many steps will it take to make a palindrome with 692 as the first addend?

To practice subtraction, have your child:

1. Write a number with four or more digits, each digit different from the oth ers. (correct: 4631 or 5297. Incorrect: 4634 or 6411); 4631
2. rearrange the digits (4631 becomes 3146); - 3146
3. subtract the smaller number from the larger ; and a. 1485
4. tell you the thousands, hundreds, and tens digit of the difference. (1,4, and 8) Tell the child that you can name the ones digit. (5) Your solution: find the total of the digits that the child has named from the difference and subtract that sum from the next multiple of 9. (9, 18, 27...) Example: In the problem above, the child will name the digits 1, 4, and 8. You will mentally add these numbers (1 + 4 + 8 = 13) and subtract the sum from 18. (18 - 13 = 5) 5 is the ones digit. (Exception: If the digits the child names total 9 or 18, the ones digit can be 9 or 0.) Tell him that you know that the ones digit of the answer is 5 but do not say how you found the answer. You will know that he has not followed directions or he has made an error in subtraction if your answer does not match the ones digit.

Other examples:

Child's Secret Number	Number Rearranged:	Problem:	Solution:
b. 5962	9265	9265 - 5962 330_ ←	(3+3=6) (9-6=3)
c. 8047	4078	8047 -4078 396_ ←	(3+9+6=18 27-18=9)
d. 8047	4087	8047 -4087 396_ ←	(3+9+6=18 18-18=0)

In examples c. and d. the sum is a multiple of 9 so the ones digit can be 9 or 0.

After the child has tried to guess how you found the secret number, ask some leading questions, such as:

1. What is the sum of all the digits of your answer? (Ex. a. 18b.9c.27d.18)
2. How are these numbers alike? (They are all multiples of 9.)
3. If I know the sum of the digits you name, how can I find the other digit of your answer? (Subtract the sum from the next multiple of 9.)

Be patient. The child will probably need several clues. After learning the trick. the child can challenge other members of the family by giving them the directions and then naming the ones digit of their answer.

Play High or Low from Chapter 11.

MULTIPLICATION: Multiplication by 1-digit multipliers, even with regrouping, presents few problems if your child has mastered the basic multiplication facts. Some difficulty may arise, however, when there is a zero in the multiplicand (lop number). Many children multiply the number they regrouped instead of multiplying zero and adding the regrouped number.

Example:

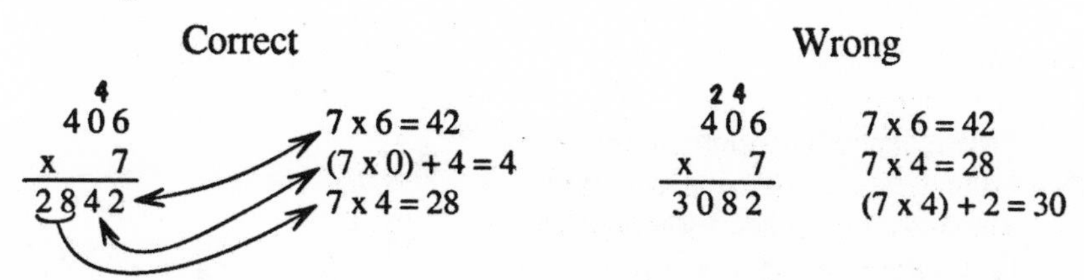

Correct
Wrong
4
406
x 7
2842
7 x 6 = 42
(7 x 0) + 4 = 4
7 x 4 = 28
24
406
x 7
3082
7 x 6 = 42
7 x 4 = 28
(7 x 4) + 2 = 30

Hint: There will never be a regrouping to the place following the zero in the multiplicand.

Examples:

```
  2 3           1  1
 20304          25006
x    8         x    3
162432          75018
```

Present situations that require multiplication of large numbers by 1 -digit multipliers. For example:

1. A space ship travels 2,142 miles in one hour. How far will it travel in 6 hours?
2. There are 5,280 feet in one mile. How many feet are in 4 miles?
3. A transport truck carries 5 cars. Each car is worth $12,362. What is the total value of the cars?
4. Uncle Joe is delivering telephone books. What is the total weight of 2.542 books if each weighs 3 pounds?
5. One minute of advertising for a special on TV costs $135,450. How much will 7 minutes cost?

DIVISION: Your child, after learning to divide 3-digit numbers by 1-digit numbers, should have little difficulty dividing 4-, 5-, or 6-digit numbers. Stress that the steps in a division problem (divide, multiply, subtract, bring down the next digit) are repeated until all digits in the dividend are used.

Explain that every division problem is made up of a series of small problems. For example: 6)34781 6)34 (quotient 5)

Read: ≤ is less than or equal to.

Think: 6 x _ ≤ 34: Six times what number (greatest number) is less than or equal to 34? (5)

Place the 5 above the 4 to show that we divided 34. not 3. Next, multiply 6 x 5; write the product under 34; subtract; compare this difference to the divisor (Continue with the process if the difference is less than the divisor); bring down the next digit-7.

Again, we just have the small problem 6)47

```
   5
6)34781
  30
   47
```

Think: 6 x __ ≤ 47; Six times what number (greatest number) is less than or equal to 47. (7)

Continue to divide, multiply, subtract, and bring-down the next digit as long as you can.

When there are no more digits to bring down, the division is complete. The 5 is a remainder. Write it as part of the quotient as R 5 or as the fraction 5/6. To check the problem, multiply the quotient by the divisor and add the remainder.

```
   5 7 9 6
6) 3 4 7 8 1
  -3 0
     4 7
    -4 2
       5 8      6 x __ ≤ 5 8 (9)
      -5 4      6 x __ ≤ 4 1 (6)
         4 1
        -3 6
           5
```

```
 5796 Quotient
x   6 Times divisor
34776
+   5 Plus remainder
34781 Equals dividend
```

A problem with a zero in the quotient may be troublesome.

Remind your child that every time a digit is brought down, a digit must be put in the quotient. In the problem on the right, that digit is zero.

```
   4607
8)36856
  32
   48
   48
     5   [8 x __ ≤ 5 (0)]
     0
     56
     56
```

If your child has difficulty completing the problem or with using all of the digits in the dividend, use graph paper or notebook paper turned sideways to help him align the digits.

	1	1	9	5	0	R 1
7)	8	3	6	5	1	
	7					
	1	3				
		7				
		6	6			
		6	3			
			3	5		
			3	5		
					1	
					0	

Sometimes children make errors in division by leaving a remainder that is greater than, or equal to. the divisor. To stress that the remainder must be less than the divisor, play Possible Remainders. Name a divisor or show a division exercise and give a point to the player that writes or names the possible remainders.

Divisor	Remainders
6	5,4,3,2,1
3	2,1

Find situations or topics for discussion that will require the child to divide large numbers by 1-digit numbers to find the answers to questions. Examples:

1. It is 238,856 miles to the moon. How fast will a spaceship have to travel to get there in 8 hours?
2. The odometer on our car reads 94,745 miles. We have had the car 7 years. If we drove the same number of miles each year, how many would that be?
3. The airplane flew 3624 miles in 6 hours. If it flew the same distance each hour, how many miles per hour did it fly?
4. Our new car cost $8975. If we pay for the car in 5 years, how much will we pay each year?

Play High or Low from Chapter 11.

ADDITION AND SUBTRACTION OF DECIMALS: Check to sec if your child follows the same procedure for adding and subtracting decimals as for adding and subtracting dollars and cents.

$ 4.23	4.23 meters	$8.15	8.15 meters
+1.79	+1.79 meters	-3.32	-3.32 meters
$6.02	6.02 meters	$4.83	4.83 meters

It is important to develop the understanding of these processes with the concrete models that you have available (dollars, dimes, pennies; meter, decimeter, centimeter measures; 10 by 10 squares of graph paper). Examples:

1.	3 dimes		3 tenths		.3
	+4 dimes	⟶	+4 tenths	⟶	-.4
	7 dimes		7 tenths		.7

2.	65 ¢ +75 ¢ 140 ¢	→	65 hundredths +75 hundredths 140 hundredths	→	.65 +.75 1.40
3.	36 cm −14 cm 22 cm	→	36 hundredths −14 hundredths 22 hundredths	→	.36 −.14 .22
4.	2 m 25 cm −1 m 9 cm 1 m 16 cm	→	2 and 25 hundredths −1 and 9 hundredths 1 and 16 hundredths	→	2.25 −1.09 1.16

5. Show a meter stick and a string or cord and say: "This cord is 8 decimeters long. What part of a meter is this? (.8) I'm going to cut 14 centimeters from the cord. What pan of a meter is that? (. 14) How can we find how much will be left?"

.8 −.14	.80 −.14 .66

   Use the meter stick to show that 8 decimeters (.8) is the same as 80 centime ters (80) and record as on the right.

From experience your child will probably conclude that the most important step is to align the decimal points. Write an exercise horizontally, such as 3.7 + .28, and ask her to write it vertically and add.

3.7 +.28	3.70 +.28 3.98

Once the child understands that 10 tenths is equal to one whole and that we regroup from tenths to ones and ones to tenths like we regroup from ones to tens and tens to ones, adding and subtracting decimals is as easy as adding and subtracting whole numbers. Stress: Write the decimal point in the answer directly under the decimal point in the problem and add or subtract the numbers.

Be alert for situations that require your child to add or subtract decimals to answer questions.

Examples:

1. "The odometer read 3,657.2 this morning. It now reads 3,892.5. How far have we traveled today?"
2. Give the heights of family members in meters (tenths and hundredths) and ask your child to find differences between the heights and the sum of all the heights.
3. This morning the baby's temperature was 102.6 . It is now 99.7. How much has it dropped?"

MULTIPLICATION AND DIVISION OF DECIMALS BY WHOLE NUMBERS: Your child has had experience multiplying and dividing decimal numbers by whole numbers when multiplying and dividing dollars and cents. Show that the same process is used with or without the dollar sign.

\$3.45	3.45	\$ 1.15	1.15
x 3	x 3	3) \$ 3.45	3) 3.45
\$10.35	10.35		

To develop understanding of the processes, use concrete material explained earlier in the chapter for modeling.

1.	4 dimes	4 tenths	.4
	x 3	x 3	x 3
	12 dimes	12 tenths	1.2
2.	6 cm	6 hundredths	.06
	x 5	x 5	x 5
	30 cm	30 hundredths	.30
3.	4 dimes	4 tenths	.4
	2)8 dimes	2)8 tenths	2).8
4.	21 cm	21 hundredths	.21
	4)84 cm	4)84 hundredths	4).84
5.	7 cents	7 hundredths	.07
	3)21 cents	3)21 hundredths	3).21

Explain that the positions to the right of the decimal point in any number are called decimal places.

Examples: 2.6 has one decimal place. 2.Qfi has two decimal places. 32.5 has one decimal place.

After multiplying decimals by a whole number, your child may discover that the product of a whole number and a decimal number has the same number of decimal places as the total number of places in the factors. Therefore, to multiply a whole number and a decimal number, he must: a) multiply as with whole numbers, b) count the number of decimal places in the factors, c) place the decimal point in the product so that there will be that number of decimal places in the answer.

Examples:

2.65	2 decimal places	.04	2 decimal places
x .2	1 decimal place	x .05	2 decimal places
.530	3 decimal places	.0020	4 decimal places

decimal number by a whole number, one must divide as with whole numbers and place the decimal point in the quotient directly above the decimal point in the dividend.

Examples:

$$7\overline{)3.5}^{\;.5} \qquad 2\overline{)4.6}^{\;2.3} \qquad 7\overline{).14}^{\;.02}$$

At this stage, your child's experience with multiplying and dividing decimals has probably been limited to working with money and, possibly, metric measures. Look for situations that will require the child to use these processes to answer questions, but always be ready to use concrete material to model the process.

Examples:

1. "I worked 42.5 hours in 5 days. If I worked the same number of hours each day, how long did I work each day?"
2. "If you earn $12.50 each week cutting grass, how much will you earn in 9 weeks?"
3. "I have a macrame cord that is 12.6 meters long. If I cut 3 equal cords from this, how long will each cord be?"
4. "I drove 148.5 miles and used 9 gallons of gasoline. How many miles per gallon did I get?"
5. "We drove our car 50.5 miles per hour for 6 hours. How far did we drive?"

We have included a simple introduction to computation with decimal numbers. Consult with your child's teacher if there is any difficulty and use concrete models to aid the child's understanding.

Adding and Subtracting Fractions with Like Denominators

$\frac{2}{3}$ → Numerators ← $\frac{3}{5}$
→ Denominators ←

Adding and subtracting fractions with the same denominators are simple processes and can be introduced as a direct extension of addition and subtraction of whole numbers.

$$\begin{array}{rl} 1 & \text{apple} \\ +\,2 & \text{apples} \\ \hline 3 & \text{apples} \end{array} \qquad \begin{array}{rl} 1 & \text{fifth} \\ +\,2 & \text{fifths} \\ \hline 3 & \text{fifths} \end{array} \qquad \begin{array}{r} 1/5 \\ +\,2/5 \\ \hline 3/5 \end{array} \qquad \frac{1}{5} + \frac{2}{5} = \frac{1+2}{5} = \frac{3}{5}$$

3 apples	3 fifths	3/5
-2 apples	- 2 fifths	2/5
1 apple	1 fifth	1/5

$$\frac{3}{5} - \frac{2}{5} = \frac{3-2}{5} = \frac{1}{5}$$

Rules:

1. To add fractions with the same denominators, add the numerators and keep the denominator.
2. To subtract fractions with the same denominators, subtract the numerators and keep the denominator.

Your child, with some understanding of fractions, will probably be able to answer intuitively questions similar to the following, even without having learned the rules. Examples:

1. "I gave you 1/6, of the pie and I gave your brother 1/6. How much of the pie have 1 given away?" (2/6)
2. "The pie is cut into 6 equal pieces, so each piece is 1/6. I gave away 2/6. How much is left for dinner?" (4/6) Ask the child to write the process that gave the answer.

$$\begin{array}{r} 1/6 \\ +\,1/6 \\ \hline 2/6 \end{array} \qquad \frac{1}{6} + \frac{1}{6} = \frac{2}{6}$$

Addition

$$\begin{array}{r} 6/6 \\ -\,2/6 \\ \hline 4/6 \end{array} \qquad \frac{6}{6} - \frac{2}{6} = \frac{4}{6}$$

Subtraction

Ask questions to see if he knows the rules. "What parts of the fraction did you add (subtract)? (numerators) What part stayed the same?" (denominators) Stress: Add (or subtract) the numerators. Keep the same denominator.
Continue to present situations that require these processes to find answers to questions.

Examples:

1. "You sleep about 1/3of each 24-hour day and you are in school about 1/3 of the day. How much of the day are you asleep or in school?" (1/3 + 1/3 =2/3)
2. "What part of the day are you awake and not in school?" (3/3 - 2/3 =1/3)
3. "I will eat 1/8 of the pizza. Daddy will eat 2/8 and your sister will eat 2/8. How much of the pizza will the three of us eat?" (1/8 + 2/8 + 2/8 = 5/8})
4. "How much will be left for you?" (8/8 - 5/8 = 3/8)
5. "We need 2/3 cup of milk for this cake and 2/3 cup for the frosting. How much milk do we need? (2/3 + 2/3= 4/3) Is there another way that you can tell me how much milk we need?" (1 1/3 c.) If he doesn't know. measure the milk and rename 4/3 as 1 1/3.
6. "I need 1 1/3 yd. red fabric and 2/3 yd. white fabric for this dress. How much fabric must I buy? Is there another name for 1 3/3 (2)

Use every opportunity to ask questions that require addition and subtraction of fractions with the same denominators. This will give practice for the processes, and it will help your child appreciate the many ways he uses fractions in his daily life outside of school.

Multiplying Fractions

Usually children do not study the multiplication of fractions until much later than the third or fourth grade, but there are occasions at home when they will use the process intuitively. It's best to let them find the answers with their own methods and then show them how to write the multiplication sentence. The process is so simple that the child may see the pattern after a few experiences.

To multiply two fractions, multiply their numerators and multiply their denominators: $1/2 \times 3/4 = {}^{1\times3}/_{2\times4} = 3/8$

To multiply a whole number and a fraction, think of the wbok number as a fraction and multiply as above: $3\times 1/4 = {}^{3/1} \times {}_{1/4} = 3/4$

Use the following situations to give your child experience multiplying fractions:

1. When you are halving a recipe, you may need to multiply a fraction by 1/2. Tell the child the recipe calls for 1/2c. of milk, but you want to make half of the recipe. Ask how much you need. Explain that taking one half of something is the same as multiplying by 1/2. Show the multiplication sentence (1/2 x 1/2 = 1/4) and ask the child to get the appropriate measure. (1/4 c.). If you must find 1/2 of 2/3 accept the answer of 1/3. If the child multi plies (1/2 x 2/3 = 2/6) discuss 1/3 as another name for 2/6. (See Chapter 5.)
2. When you are doubling a recipe, you will probably need to multiply a frac tion by a whole number. Tell your child that the recipe calls for 2/3 c. of milk and ask how much you need if you double the recipe. He can add (2/3 + 2/3) or multiply by 2. Show how to write the multiplication sentence.
   $2 \times 2/3 = 2/1 \times 2/3 = {}^{2\times2}/_{1\times3} = 4/3 = 1\ 2/3$
   Let the child use the measuring cups to find: 4/3 = 1 1/3.
3. Show a pie cut into thirds and say: "Each piece is 1/3. If I give you 1/2 of a piece, what part of the pie will you get?" Score each piece to show 1/2 of 1/3. Ask the child to count the pieces (6) and name each small piece, (1/6) Show the multiplication sentence: $1/2 \times 1/3 + {}^{1\times2}/_{2\times3} = 1/6$

$\frac{1}{6}$ $\frac{1}{3}$

Use other opportunities to show your child how to work with fractions. Demonstrate the process with models when necessary.

Solving Two-Step Problems

Throughout this book, the emphasis has been placed on problem-solving, the most relevant aspect of mathematics. Your child, because of having developed many computational processes, should be ready to solve more complex problems. Often these problems require more than one, or even two steps. To motivate problem-solving, use topics of greatest interest to your child and situations that occur most often in his daily life.

PROBLEM-SOLVING AIDS: Use the appropriate aids from the list below if your child has trouble solving problems that you present.

1. Check to see if the child has the necessary computational skills to solve the problem.
2. Encourage the child to "act out" the problem and tell the appropriate com putational processes for solving the problem. Example: Problem: "This morning you had $12.45. You spent $3.15 and earned $2.50. How much do you have now?" Display $12.45 and ask the child to show how the amount of money changed. Discuss the computational processes. "Did you take away any money? What process will you use to show this? (Subtraction.) Did you earn some of the money back? How will you show this?" (Addition.)
3. When problems involve very large numbers, sometimes it helps to substitute small numbers, solve the problem and then follow the same steps with the original numbers.
4. Encourage your child to look for unnecessary information in a problem and simplify the problem by marking out this extra information. Problem: "Kim is putting tomatoes and onions in her garden. She put 20 onions in each of 4 rows and 16 tomato plants in each of 3 rows. How many onions did she plant?" (Child should mark out "and 16 tomato plants in each of three rows.")
5. Remind the child to be sure that there is sufficient information to solve a problem. Problem: "Carol paid $3.00 for a notebook, 65< for a pen, and 15< for a lead pencil. How much change did the clerk give Carol?" What is need ed? (How much money did Carol give the clerk?)

FINDING AVERAGES: Give some marbles (any small items) to your child and two other people - 7 to one, 3 to one, and 2 to one. Ask: "What is the average number of marbles that I gave to each person?" If the child does not know the meaning of the term average, say: "If the three of you share these marbles evenly, each getting the same number, how many will each get?" Join the sets of marbles (find the sum of the numbers), then redistribute the marbles evenly into 3 equivalent sets. (Divide the total by 3.) Explain that this number (4) is the average number given to each person. Repeat the activity with other items and other numbers. Stress: To find an average of a set of numbers, add the numbers and divide the sum by the number of addends.

Suggested activities:

1. Ask the child to keep a record of the temperatures each day for 7 days at the same hour. Ask: "What was the average temperature at that hour for the seven days?" (Add temperatures; divide by 7.)
2. Encourage the child to keep a record of his test scores for 5 tests and to find the average score.
3. If the child is involved in any athletics, ask for the teams scores for six games and the average score for each game. Remind the child that if the team fails to score in any game, a zero must be recorded and counted as an addend to find the number by which the sum is to be divided.
4. Talk about records of the child's favorite athletes and how their averages are figured.
5. Keep a record of the length of time it takes the child to make the bed and straighten his room each day for 6 days. Ask for the average number of minutes.

6. Ask the child to find the weight (height) of each family member and to find the average weight (height) of the members of the family.

Finding Perimeters of Polygons

Activities in Chapter 5 introduced methods for finding the perimeters (distances around) of squares and other rectangles. Your child can find the perimeters of these and other polygons by adding the lengths of their sides if he can add a column of more than two addends. Watch for these trouble spots and make the following suggestions.

1. He does not list the length of all sides of the shape. Suggestion: Count the addends Incorrect. in the column, the sides of the polygon. and compare the two numbers. They should be the same.
2. He cannot find the sum of measures given in fractional parts of a unit. Suggestion: List the length of each side to the nearest whole unit or review addition of fractions.

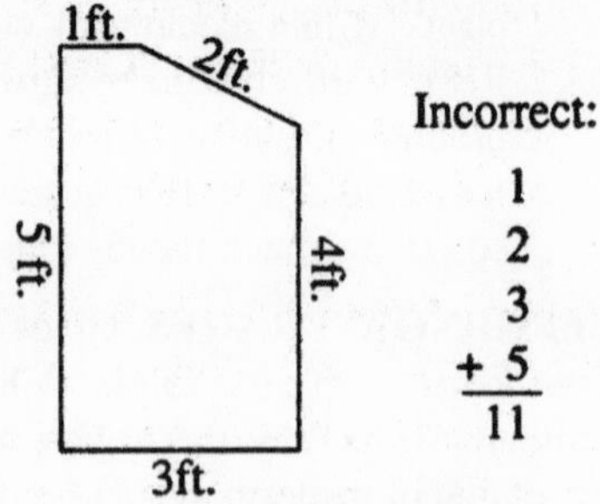

3. He measures the sides of a polygon but lists the measures in different units. Suggestion: Write the name of the unit beside each addend and stress that all units must be the same. Example: Change all units in the figure on the right to feet

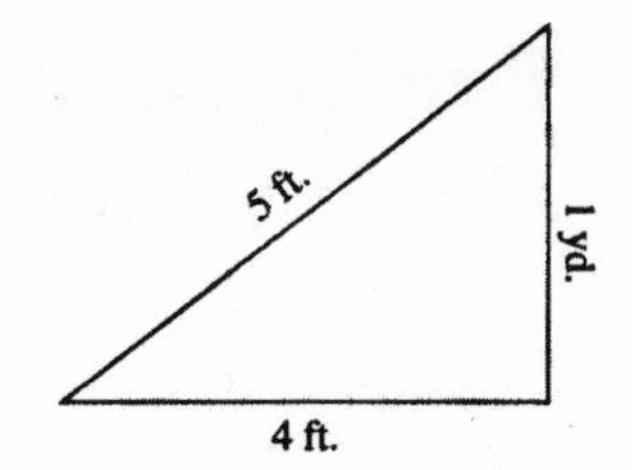

Allow your child to use the following shortcut, but do not stress it as a method at this point. If two or more sides of a polygon are the same length, multiply that length by the number of those sides, and use the product as an addend to find the perimeter of the polygon.

In the figure on the right, two sides are 5 ft. long, three sides are 3 ft. long, and three sides are I ft. long.

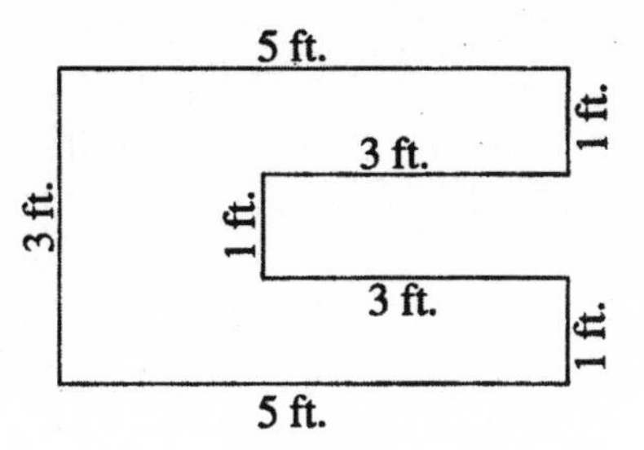

(2x5) + (3x3) + (3x1)
10 + 9 + 3 =22

Suggested activities:

1. Have your child find the perimeter of a room. porch, deck, patio, or any other part of the house that is a polygon other than a rectangle. Discuss the unit of measure that is the most appropriate to use (inch, foot, yard; centime ter, decimeter, meter).
2. Have the child help you find the perimeter of a yard or lot that is a polygon other than a rectangle.
3. Ask the child to draw on a sheet of graph paper as many rectangles as he can with a perimeter of 12 units.

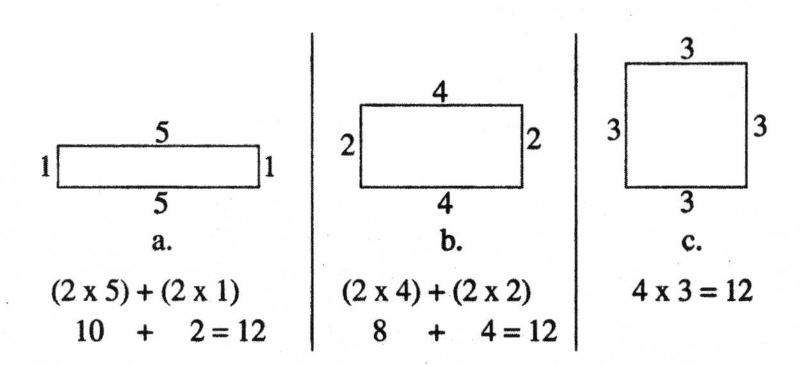

Finding Areas of Rectangles and Other Shapes

A method for finding the areas of rectangles is shown in Chapter 5. Review this method by asking the child to find the area of each rectangle drawn on graph paper with a perimeter of 12 units.

a) 1x5=5 b)2x4=8 c)3x3=9

If necessary, use the activities in Chapter 5 to review:

length x width = area of a rectangle

Draw or shade shapes similar to the following on graph paper and ask the child to find the area by counting the squares in each space.

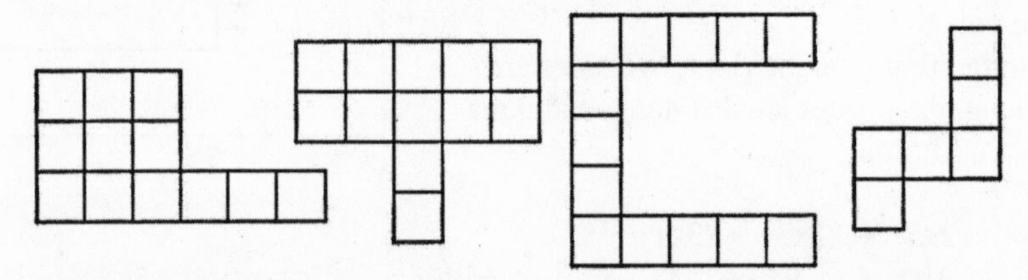

After many experiences, the child may see small rectangles within the shapes, multiply to find the area of each, and then add to find the area of the shape.

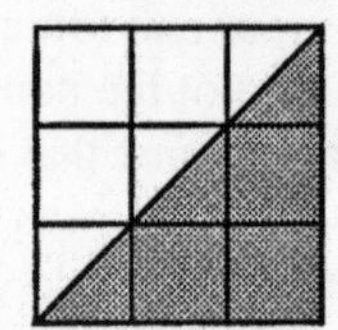

Draw a rectangle and shade a triangle as shown on the right.

Remember! *A square is a kind of rectangle.*

Ask: "What is the area of the square? (9 square units) The triangle? (41/2 sq. units) How do you know?" (The triangle is half of the rectangle and 1/2 x 9 = 4 1/2) Area of a triangle =1/2 x base x height

Finding Volume

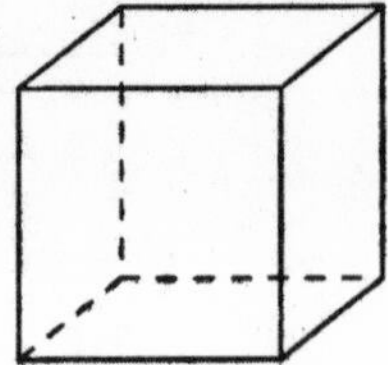

Show your child physical examples of cubes (a child's block, a die, some boxes), identify them as cubes, and discuss how they are alike: "How many edges does each cube have? (12) Are all of the edges on one cube the same length? (Yes) How many faces does each cube have? (6) What shape are all of the faces?" (Square)

Show your child a box and say that you would like to find the \-iiliiiw of the box. Explain thai volume tells ihe number of cubes it lakes 10 till ihe space inside. Explain that cubic units are used to measure volume just as square units are used to measure area. Encourage him to trace the pattern below, cut along the solid lines, fold on the dotted lines, and paste the tabs to tbnn a cube wnh a volume of one cubic inch.

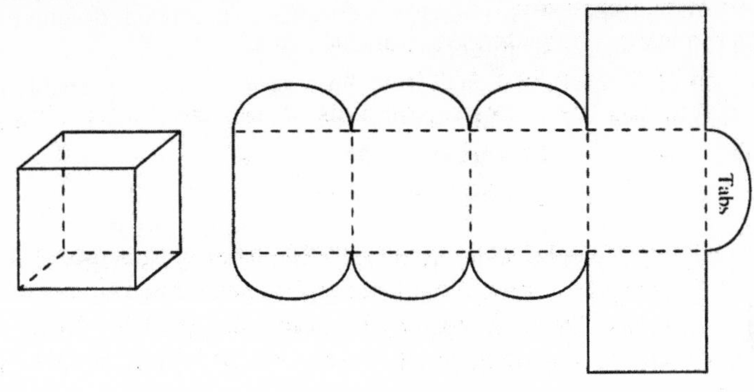

If you have cubes that til exactly into a box (children's blocks), ask: "How many cubes can you put in one row in the bottom of the box? Try it. How many rows can you put in the bottom? How many cubes did you put in this layer? How many layers can you put in the box? How many cubes?" Explain that the number of cubes tells the volume of the box.

Do not stress the formula (length x width x height) for finding volume at this point. With experiences of this type. the child may discover it. If you do not have material for the above activity, encourage the child to use the cube made of paper or tagboard when estimating how many cubic inches it will lake to till small boxes. Follow the same procedure with cubic centimeters as with cubic inches.

Discuss a cubic fool and a cubic yard. Encourage your child to build either or both of these and estimate the number it will take to till a room.

Miles and Kilometers

Your child has been measuring with inches, feel, and yards and probably knows the relationship between the measures.

1 inch (in.)= ____________

1 foot (ft.) = 12 in.

1 yard (yd.) = 3 ft. or 36 in.

Most children have heard distances expressed in miles and have some feeling for the length of a mile, particularly if they jog or run track in school. Ask your child how many feet are in one mile. (5,280) Give distances of one to nine miles and ask for the same distance in feet. "It is 2 miles to your school. How many feet is that?" (2 x 5280 = 10,560)

Ask for the number of yards in one mile. (5280 +3=1,760) Ask which unit of measure would be best for measuring these distances: a)
Distance to the library; b) his height; c) distance to the moon; d) length of a fish; e) length of his bike; 0 length of a hallway, etc.

If your child has been using metric units of measure, he probably knows these relationships: 1 centimeter (cm) ,_____
1 meter (m) = 100 cm

Many children remember hearing the term kilometer used in races (Olympic Games, etc.). Some speed signs are marked in miles and kilometers per hour and some automobiles have the speedometer labeled in miles and kilometers.

Discuss the length of a kilometer: 1 kilometer (km) = 1000 m. One kilometer is also a little more than !A mile.

Ask: "If one kilometer equals 1000 meters, how many meters does 2 kilometers equal? (2000) It is four kilometers from here to your school. How many meters is that?" (4000)

As with inches, feet, yards, and miles, ask which unit-centimeter, meter or kilometer-is most appropriate for measuring various distances: a) Length of the child's foot; b) length of a kite string; c) perimeter of a book; d) perimeter of a room; e) distance to the park; or 0 distance to Grandma's house, etc.

Scale Drawings

Your third- or fourth-grade child may be learning to make and read scale drawings. He may have encountered the term scale when working with model cars, airplanes, or doll houses and furniture. When you and your child are making or reading scale drawings, stress these points:

1. Drawing to scale results in a figure that is the same shape as the original but different in size.
2. The choice of scale is arbitrary: 'A inch may stand for 5 yards in one draw ing and 10 yards in another.

To simplify scale drawing, begin with graph paper and let the side of a square stand for a given measure. Example: Child's bedroom: 9 ft by 12 ft. = 3 yds. by 4yds.

In Drawing A, the side of a square equals 1 foot.
In Drawing B, the side of a square equals 1 yard.
The shapes are the same but the sizes are different.

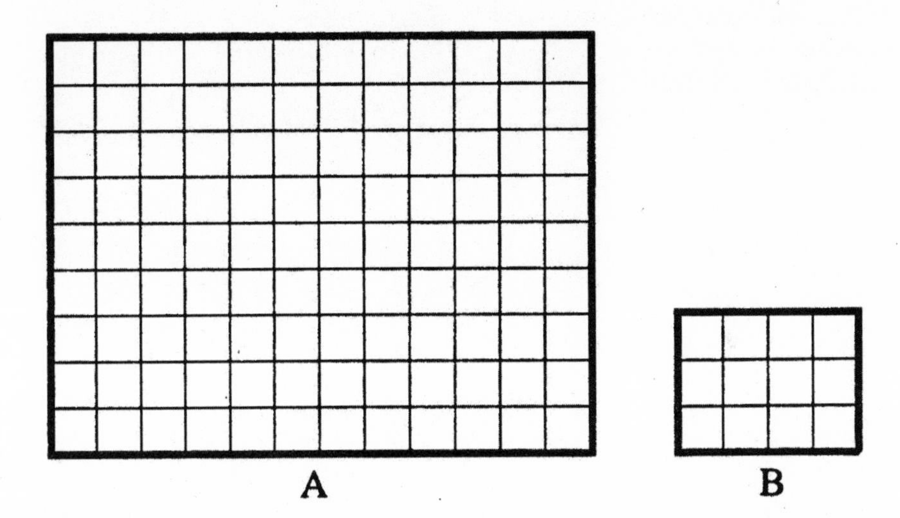

Encourage your child to make scale drawings of rooms, city blocks, furniture, etc.

Summary

The skills in Chapter 6 may not be encountered by your child until he is in the intermediate grades, four to six. These skills are extensions of those in Chapters 1 to 5 and should be presented to the child only after he has mastered the previous skills.

PART 2

MONEY: EARNING IT, SPENDING IT, AND SAVING IT 7

(a) 499

(b) Jan. 6 20 00

(c) Pay to the order of Samuel Gladney (d) $ 46 25/100

(e) Forty-six and 25/100 Dollars

All Savers' Bank

(f) Sarah Hutz

Probably no application of mathematics will have more meaning to your child than the use of money. When he becomes aware of its value for getting other items, his interest in learning to handle these coins and bills efficiently will be kindled. This interest will lead to opportunities for the child to practice counting, computation, and problem solving. It will also provide opportunities for you to introduce new math skills and concepts when they are most relevant to him. The bills and coins are excellent objects with which to model math problems. Therefore, the purposes of this chapter are many. The activities described here arc designed to help your child keep the value of money in perspective and to learn to handle money with skill. They will also use the child's interest in handling money to improve his math skills. Adjust the level of difficulty of these activities to fit the skills and interests of your child.

Earning It

Though each member of your family probably has regular duties around the home that match his age and ability, chores often arise for which no one is responsible. Your offer to pay young family members to perform these chores may provide their earliest experiences in earning money. These opportunities to cam money will be accompanied by responsibilities: the responsibility to make wise decisions that involve money and the responsibility to count and compute accurately. Guide your child as he meets these new responsibilities.

CHECKING INCOME: Encourage your child to count the money when he has been paid, for a chore. The first time that he discovers he has been underpaid he will begin to realize the importance of counting accurately to check all money transactions. You can control the level of difficulty of the counting activity with the size of the coins or bills that you use for payment. If he is learning to count by ones, pay him with pennies; if he is learning to count by fives, pay him with nickels, etc.

RECORDING INCOME: When your child has learned to add with regrouping, or carrying, encourage him to keep a written record of his income (earnings and allowance) for a few days at a time. This activity will present meaningful addition practice for the child and it will also reinforce the method of recording money numbers with decimals. As he writes the money numbers, stress the importance of using the decimal point to separate dollars and cents. Discuss the difference between pricing an item at $4.85 rather than $485, the same digits without the decimal point.

Stress: Place the decimal points under each other when writing the numbers vertically for addition or subtraction.

Examples:

$5.60	$8.16	$.04	$9.27
+ 10.06	-0.47	+ 6.00	-8.00
$15.66	$7.69	$6.04	$1.27

Suggest that he use a form similar to the following:

Record of Income for June 1 – June 7, 1991:					
Earned:	June 1:	$ .45		Balance:	$ .77
Earned:	June 2:	+ .15	Earned:	June 5:	+ .20
	Total:	$ .60		Total:	$ .97
Earned:	June 3:	+ .17	Earned:	June 7:	+ 3.16
	Total:	$ .77		Total:	$ 4.13

If your child has difficulty with the addition, have him join the sets of like coins (pennies with pennies, dimes with dimes, etc.), then exchange each set of ten for the coin or bill of equal value, and record the results.

Example: 26¢
+ 17¢

Dimes	Pennies
(1)	
2	6
+1	7
	3

a. He will join 6 pennies with 7 pennies, exchange 10 pennies for a dime, and record the results as on the right:

Dimes	Pennies
(1)	
2	6
+1	7
4	3

b. He will count the dimes and record the total number as is shown on the right.

Choosing a Method of Payment: As your child matures and begins to earn money from sources outside the family, he will probably encounter various methods of payment, such as: a) By the job; b) by the hour; and c) by the day.

Offer these methods to him in the home to prepare him for making competent decisions in the future. Before making the child an offer, find out if he has the math skills that will be involved in the computation. If he has not learned to multiply a money number by a whole number, encourage him to model the problem with dollar bills and coins. The results may be recorded as multiplication or as repeated addition.

Example: $ 1.45 x 3 or $ 1.45 + 1.45 + 1.45

a. He will join 3 sets of 5 pennies (15), exchange 10 for a dime, and record the results as shown on the right: $ 1.45 x 3 = 5 (1 carried)

b. Next he will join 3 sets of 4 dimes (12), add the dime that was exchanged for the pennies (13), trade 10 dimes for a dollar, and record as shown on the right. $ 1.45 x 3 = 35 (1 carried)

c. The child will now join 3 sets of 1 dollar (3), add the dollar that was traded for the dimes (4), and record as shown on the right. $ 1.45 x 3 = $ 4.35

Offer two methods of payment for the same job and encourage the child to estimate the time involved and choose the method he prefers.

Example: Offers for cleaning (or helping to clean) the garage:

a. $4.00 for the job, or
b. $1.50 per hour.

Guide the child to see that he needs to:

1. Estimate the time needed to complete the job;
2. multiply the amount offered per hour in offer b above ($1.50) by the number of estimated hours; and
3. compare the results in step 2 to the amount offered in offer "a."

Estimated Time in hours:	times	Amount per hour:		Total amount:
2	x	$1.50	=	$ 3.00

Four dollars named in offer a is greater than three dollars, the result of offer b, so offer a seems to be better.

Suppose the job actually takes 3 hours.

Actual Time in hours:	times	Amount per hour:		Total amount:
3	x	$1.50	=	$ 4.50

In this instance, the offer of $1.50 per hour in offer b would be better than offer a.

BIDDING ON A JOB: The competition of the job market often requires a person to bid on a job by telling how much he will charge to complete the task. The employer asks other people to name their charges at the same time and the person who names the lowest price, all else being equal, is given the job. The ability to make reasonable bids requires the skills of estimating the time required for the job, the difficulty of the job, and any expenses that might be involved. You can provide practice for developing these skills by asking for bids on jobs around the home from family members, or from your child and other young members of the community.

Examples:

1. Mowing the lawn. Encourage the child to:
   a. Determine what is expected (Sweeping? Trimming? Whose equipment will be used?);
   b. estimate the length of time it will take to complete the tasks; and
   c. estimate the cost of fuel and wear and tear on the equipment required for the job. The total expenses must be subtracted from the amount bid

   to determine the income for the bidder.
2. Babysitting with younger members of the family. Ask the child to bid on this job by naming an amount per hour. When you tell him the approximate number of hours, encourage him to multiply to find the total amount of money.
3. Cleaning. Although the child will bid this task by the job, the time required for completing the chore must be considered carefully.

At first, the inexperienced bidder may be inclined to try to underbid his competitor. After several experiences, some in which he has found herself being grossly underpaid, he will probably begin to make more reasonable bids on the jobs that are offered.

COMPARING JOBS: Another skill that your child will hopefully need when he enters the job market of adult life is that of comparing jobs to determine which he will accept. Many factors will influence his final decision and one of these is the income he will receive. You can prepare him for this decision-making experience by providing similar situations in the home.

Examples:

1. Tell your child that you need a babysitter for three hours at $2.00 per hour and you also need someone to mow the grass for $6.00. Encourage him to estimate the length of time needed to mow the grass and to multiply to find the total amount that he can earn by babysitting. In this case, she will probably choose the job that requires less time.
2. Offer the job of weeding the flower beds for $2.50 or running the sweeper and dusting for $3.00. Encourage the child to estimate the amount of time each job will take and divide to find which job will pay more per hour. Encourage your child to use bills and coins to model this division.
   Example:

   a. He will divide 2 dollars into two equivalent sets ($1 in each set) and record.

   ```
        1
   2)$ 2.50
      2
   ```

   b. Next he will divide 5 dimes into two equivalent sets (2 in each with 1 dime left over) and record.

   ```
        1 2
   2)$ 2.50
      2
        5
        4
        1
   ```

   c. Now he will have to exchange the remaining dime for ten pennies, divide these into two equivalent sets and record.

   ```
     $ 1.25
   2)$ 2.50
      2
        5
        4
        10
        10
   ```

   Weeding: 2 hrs. Cleaning: 3 hrs.

   ```
     $ 1.25        $ 1.00
   2)$ 2.50      3)$ 3.00
   ```

Offer other choices between two jobs that require the child to estimate the time and compute the income to determine the most desirable choice.

Spending It

Your child's spending habits and attitudes will be greatly influenced by those of the adults around him. The child's shopping skills should be developed under supervision. The following activities will help develop these skills as well as provide meaningful practice in counting, computation, and problem solving.

Using Newspaper Ads and Mail Order Catalogs: These are great tools to use for preparing your child to be a wise and skillful shopper. Activities involving the use of these materials will motivate him to use his math skills and problem-solving techniques. They also offer information that will help your child make more intelligent buying decisions.

Examples:

1. To help your child learn to read prices and model each, suggest that he find an item in the newspaper or catalog that he would like to buy. Encourage him to read the price and show the amount of money needed to buy the item.
2. For practice of addition of money numbers, ask your child to :
   a) Find two items that he would like to buy;
   b) add to find the total cost of the items; and
   c) show you the amount of money that he needs to buy the items.
3. To provide experience with subtraction of money numbers, ask your child to choose one item that he would like to buy. Show him a bill or set of coins of greater value than the cost of the item, and ask him to find how much change he will receive when he pays for the item, with that bill or set of coins. Encourage him to use coins and bills to model the subtraction, if necessary.

   Example: Cost of pencil: 16¢. Bill used: $1.00

   a. He will exchange one dollar for 10 dimes and record.

   $$\begin{array}{r} \textcircled{10} \\ \$\cancel{1}.\cancel{0}0 \\ -\ .16 \\ \hline \end{array}$$

   b. Next he will trade one dime for 10 pennies and record.

   $$\begin{array}{r} \textcircled{9} \\ \cancel{\textcircled{10}}\textcircled{10} \\ \$\cancel{1}.\cancel{0}0 \\ -\ .16 \\ \hline \end{array}$$

   c. He will remove 6 pennies and record the number of pennies left (4).

   $$\begin{array}{r} 9 \\ \cancel{\textcircled{10}}\textcircled{10} \\ \$\cancel{1}.\cancel{0}0 \\ -\ .16 \\ \hline 4 \end{array}$$

   d. Now he will remove 1 dime and record the number of dimes left (8).

   $$\begin{array}{r} 9 \\ \cancel{10}\textcircled{10} \\ \$\cancel{1}.00 \\ -\ .16 \\ \hline .84 \end{array}$$

4. For practice of multiplication of money numbers, have your child choose an item to buy. Decide on how many of the items you will let him buy, and ask him to find how much they will cost.
   a) 4 pencils at 16¢ each.
   b) 3 puzzles at $1.95 each.
5. For practice of division of money numbers, suggest that your child choose an item and find how many he will be able to buy with a given amount of money.
   a) How many balloons at 6¢ each will you buy with 45¢?
   b) How many pencils at 16¢ each will you buy with 79¢?

LEARNING WITH GAMES: Teachers often involve children in games that provide practice for handling money. "Play" money is used for convenience in the classroom. but children are more motivated when they handle real money. You have the advantage of being able to provide real money at home.

The game of Fewest Coins in Chapter 11 provides practice for counting skills. Playing the game also helps develop skills of counting with coins; skills that are necessary for making or counting change.

Adjust the game of Store described in Chapter 11 for practice of counting and computation skills and the skill of making change. Adjust the game by changing:

a) The cost of the items to be bought or sold;
b) the size of the coins or bills used to make purchases;
c) he number of items in each sale; and
d) the difficulty of the questions asked by you about each sale or purchase.

SHOPPING: As soon as your child begins to identify coins, let him participate in some shopping activities. Show prices on items and encourage (he child to count out the amount of money needed to make the purchase. As he learns to add. subtract, multiply, and divide, plan shopping experiences that involve these processes. Do not expect him to perform mental computation with processes that he has only recently learned. Have paper and pencil available and be patient as he finds answers to questions similar to the following:

Addition: Do you have enough money to buy the box of crayons for 40< and the coloring book for 59<? (Help him And the amount of tax on the purchase.)

Subtraction: If you give the clerk a dollar bill when you buy the paints for 69e, how much change will you get? (Urge him to count the change after each purchase.)

Multiplication: Each bar of candy costs 23<. How much will you have to pay to buy one for each of four people?

Division: How many 7(sticks of candy can you buy with your quarter?

If your child cannot complete the computation to find the answers to your questions, ask how he would find the answers-which process he would use-addition, subtraction, multiplication, or division. Remember, he may have a handheld calculator but he still needs to know which process to use.

COMPARING BUYS: In order to be a wise shopper, your child needs to develop the skill of comparison shopping. Comparison shopping helps to answer the question: "Which is the better buy?"

Although the involvement of your primary child in this shopping activity will be determined by the limitations of his computation skills, he will profit by observing and discussing your decision-making techniques. He will probably become aware of your reasons for buying "economy" size containers at the grocery store and may assume that it is always more profitable to buy the larger container. This is not really true.

These cans of frozen apple juice concentrate were. recently seen in the frozen-food department of a grocery store:

6 oz. cans at 45¢ per can
12 oz. cans at 95¢ per can

Two small cans contain the same amount of juice as a larger can and will cost 90¢, or 5¢ less than the larger can. So the "economy" size does not always represent true economy.

Encourage your child to practice comparison shopping by involving him in activities similar to the following:

1. Let the child plan a luncheon or other get-together for friends and assist you with shopping. When choosing between two brands of food or two sizes of containers, discuss the better buy with him.
2. Suggest that he assist with planning and shopping for holiday meals and other festivities, always stressing the importance of finding the better buy.
3. Ask for his assistance with planning and shopping for a camping trip or pic nic and with selecting the better buys.

Provide as much assistance with computation as is necessary to determine the better buys in these activities.

SHOPPING AT SALES: Learning to buy "on-sale" items wisely requires careful shopping. Train your child to watch closely for flaws in merchandise on sale. Alert him to the fact that often items bought on sale cannot be returned, so it is extremely important to be assured of the correct size and quality of the merchandise.

Your primary child probably has not learned the concept of percent, but if you are involving him in shopping activities he may have encountered such terms as:

5 percent sales tax
30% off on all items

When he shows curiosity about the meaning of these terms, explain them simply and show other ways of writing percent:

5 percent = 5 % = .05 = 5/100 (Means 5 out of 100)
30 percent = 30% = .30 = 30/100 (Means 30 out of 100)

Examples:

1. Five percent sales tax means that 5¢ must be added to the cost for every dollar. A toy costs $6.00 and there is 5% sales tax:

   $$6 \times 5 = 30$$

   $$\$6.00 + \$0.30 = \$6.30$$

   $6.30 must be paid for the toy.
2. Thiry percent off on all items means that the regular price will be reduced 30¢ for every dollar. A toy that regularly sells for $9.00 is on sale with 30% off.

   $$9 \times 30¢ = \$2.70$$

   $$\$9.00 - \$2.70 = \$6.30$$

   The sale price of the toy is $6.30.

Making Change by the "Drugstore Method": When your child makes a purchase, the clerk will probably count his change by beginning with the amount of the purchase and continuing to the amount of the coin or bill used for payment. It is important that he understands this method in order to count with the clerk or to check his change rapidly before leaving the counter.

Example:

1. Cost of item: Coin or bill used for payment:
   13¢ $1.00

To count the change, begin with the cost of the item and count pennies to the next number that ends in 0 or 5. ("Thirteen, fourteen, fifteen.") Then use coins of the largest value possible to reach the value of the next coin or bill.

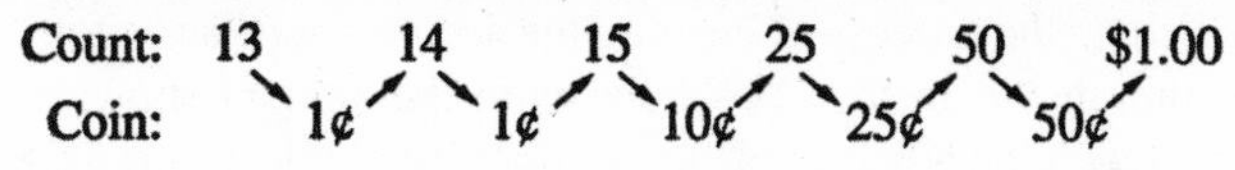

2. Cost of item: Coin or bill used for payment:
   $3.64 $5.00

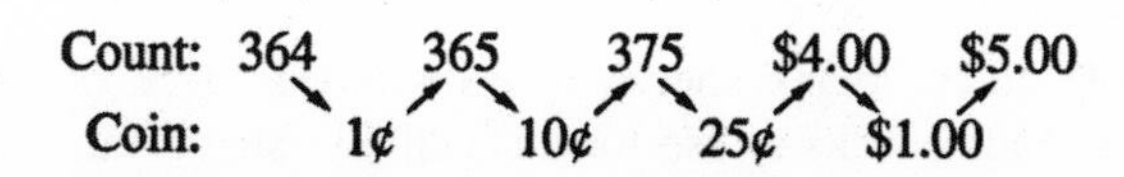

Practice this method of counting change at home with role playing. Give the child a coin or bill ($10.00), and suggest that he select from a catalog items that he would like to buy for less than that amount. When he gives you, the clerk, the coin or bill, count the change as you give it to him using this method.

Encourage him to count with you to check for possible mistakes.
Reverse the roles. You choose the item and have him count the change back to you.

BUYING BY MAIL ORDER: When you order merchandise, ask your child to complete the form under your supervision. He will add to find the total and he will multiply to find the amount of tax due on the order.

RECORDING INCOME AND EXPENDITURES: As long as your child is keeping his money in his own bank or his pocket, a written record of his earnings, how much he has spent, and his balance will be helpful. This record can follow the same pattern as the record of income described on a previous page.

Record of Money Earned and spent for June 1 – June 7, 1991

June 1:	Earned:	$ 5.00	Balance brought forward:		$ 3.92
	Spent:	1.79	June 5:	Spent:	2.29
	Balance:	$ 3.21		Balance:	$ 1.63
June 2:	Spent:	.43	June 7:	Earned:	$ 5.00
	Balance:	$ 2.78		Balance:	$ 6.63
June 3:	Earned:	$ 1.14			
	Balance:	$ 3.92			

THE BANKING GAME: Checking accounts present challenges to some adults. These challenges include the process of opening the account, writing checks. and balancing the amount shown in the checkbook with the amount shown in the bank statement. By engaging your third- or fourth-grade child in the Banking Game. you can help prepare him to meet these challenges, encourage him to keep a record of his expenditures, provide him with relevant computation practice. and increase his awareness of the importance of mathematics in his life outside of school.

Write "non-negotiable"on the blank checks of a discarded checkbook or make copies of some "pretend" checks to use for this activity. Suggest to your child that he let you be the banker for his money for awhile so that he can learn to withdraw money by writing checks and to balance a checkbook. Explain that when the two of you agree for him to buy an item, he must write a check and give it to you. the banker, before you will pay for the item with his money. Stress that you will honor checks only if he has sufficient funds in his account.

After the child has deposited his money with you, discuss the different parts of a check and the purpose of each pan.

For discussion, begin at the top of the check and stress the following points:

a. The check number To help you identify the specific check at a later time.

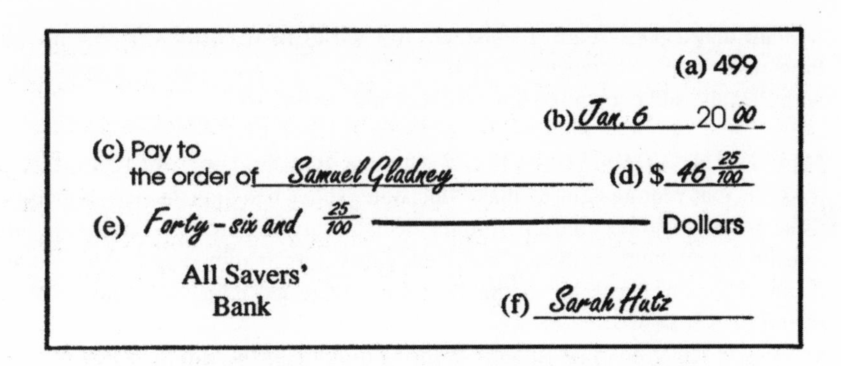

b. The date: The month, day, and year that you are writing the check.
c. Pay to the order of: the name of the person to whom you are writing the check and the person who will receive the amount of cash you specify.
d. The amount of the check: written with Arabic Numerals. This tells the amount that will be received by the person to whom the check is written. Write the dollar amount with a whole number and the number of cents as a fraction with a denominator of 100. If the check is for an exact number of dollars, write two zeros in the numerator of the fraction ($\$46^{00}/_{100}$) to prevent someone from altering the amount of the check. If the amount of the check is for a number of dollars and less than ten cents, place a zero before the number of cents for the same reason. ($\$46^{05}/_{100}$)
e. The amount of the check written out: Write the dollar amount with words and the cents as a fraction of a dollar. Writing the amount of the check in two forms also helps prevent someone from changing the amount of the check. A long wavy line here is an added protection against possible alterations.
f. Last line: your signature.

Encourage your child to write the checks and keep the records under your guidance until he feels comfortable with the process. Stress the importance of accurate computation.

Show the check stub or other part of the checkbook that is designed for the writer's personal record.

All of the information on the check that Sarah wrote to Samuel is on this check stub. In addition, the purpose of the check is written below Samuel's name. In the future, Sarah can refer to this record to remind herself of why she gave this check to Samuel Gladney. It also shows the balance in her account.

499 $ 46 25/100
1/6 19 97
To: Samuel Gladney
Rental for tent

Bal. For'd	37.16
Deposits	16.32
	3.27
Total	56.75
This check	46.25
Bal. For'd	10.50

Explain that a bank sends a statement to each of its depositors once a month. Prepare a simple statement of your child's account, showing all deposits and withdrawals, and encourage him to check his records with yours.

MISCELLANEOUS ACTIVITIES: Involve your child in as many projects as possible that require him to make decisions about handling money. Supervise these projects and become a part of the decision-making process. Encourage him to estimate amounts of money involved, gather information, and make accurate computations to check his estimates. Some suggested projects for his involvement are as follows:

1. Allow the child to assist with selecting his clothes that are to be bought with a predetermined amount of money or from his own earnings.
2. Allow the child to assist with some of the decorating of his own room by helping to figure the cost of fabric for window treatments, the amount of paint for walls, and/or carpet for the floors, and comparing prices of differ ent choices.
3. Plan a small construction project (table, dog house, bird house) with your child and ask him to help you figure the cost of the lumber, nails, and other materials.
4. Plan a camping trip, picnic, or other outing and ask his help with figuring the cost of food and other supplies.
5. Encourage the child to keep a record of his recreation costs for a month. (How much for the movies, ballgames, sports equipment, etc.?)
   Choose activities that match your child's math skills and interests. Provide as much help as is needed for him to complete each project successfully.

Saving It

From the moment a child begins to receive an allowance or earn some money, he should have the responsibility for some savings even though it is very small. In the beginning, it may be only the amount necessary for buying lunches for a few days.

PLANNING FOR SAVING: Children often want to buy items that cost more than they receive from one allowance or payment for one chore. This is an appropriate time to guide them to the realization of the advantages of saving money. At first, the goal must be small enough for the child to reach in a short time, possibly two weeks. As he matures, these goals will gradually increase and the time span should lengthen.

If you give your child an allowance regularly, encourage him to plan his expenditures and savings. Guide the child's planning by asking such questions as the following:

> What items do you need to buy with this allowance? How much will these items cost? How much will you have left after buying these necessities? How much of this do you want to save? How much will you have left after putting this amount into savings? Will this amount be enough to last until you get your next allowance or earnings? If not, what can you change?

PIGGY BANK SAVINGS: For short-term savings, provide a bank that can be opened easily so that the child can remove the money and count it at frequent intervals. Not only will this give the child a sense of security for the safety of his money, but it will also provide opportunities for him to practice counting skills.

USING A BANK OR CREDIT UNION: If your child gets a regular job or has access to a larger income, he may wish to open a savings account in a bank or credit union. Even though you will need to manage the transactions, involve him in discussions, the process of making decisions, and the computations that relate to his account. Explain the advantage of letting his money earn money by collecting interest paid by the savings institution. Encourage him to develop a pattern for saving by depositing an amount at regular intervals. This amount should be small enough to be realistic in relation to his other needs.

Take the child with you to open this account and encourage him to listen to your inquiries.

> What is the rate of interest paid on the money in the account? How often is it compounded (figured and added to the account)? Is there a penalty for early withdrawal of funds? If there are two co-signers, can a withdrawal be made without the signature of both?

Summary

The interest a child has in the value of money can be directed to improve his mathematical skills. These improved skills can, in turn, be directed to improve his ability to earn, spend, and save money.

When a child has the opportunity to earn money, he must make decisions and solve problems that affect the amount he will earn. Some of the skills that can be promoted in the home under your supervision include the following:

1. Computing;

2. choosing between methods of payment for performing a job;
3. comparing the amount of pay to be received from different jobs; and
4. making reasonable bids on a job.

Your child's ability to spend money wisely will be greatly influenced by his experiences at home. Newspaper ads, mail-order catalogs, and games are tools that can be used to develop the mathematical skill that is needed.

Closely supervised shopping experiences will provide practice for the following skills:

1. Comparing prices and values;
2. shopping at sales;
3. counting change; and
4. ordering merchandise by mail.

Some adults appear insecure about opening and using a checking account, but many children appear to enjoy the experience when it is done under the supervision of an adult. You can begin that supervision early by involving your child in The Banking Game.

Your child will profit from guidance as he attempts to develop a reasonable balance between spending and saving his money. The habit of saving must be developed gradually, beginning with small, short-term goals that are slowly enlarged and lengthened.

As your child progresses through school, the activities described in this chapter can be continued and expanded to levels that involve more advanced mathematics. Hopefully, the money experiences that you provide your child will improve mathematical skills, promote good decision-making techniques, and develop a balanced attitude toward the value of money-the best ways to earn it, when to spend it, and when to save it.

8
MATH AND THE NEWSPAPER

20¢ Off

Gurgle-Ade
Makes 6 Quarts

Limit: 1 coupon per Item
Good Through 7/11/91

You have, at your fingertips, one of the most meaningful and current learning resources available-the newspaper. With a little guidance, your child can use this source of knowledge to satisfy his curiosity about the world. Strong math skills will help the child interpret information, and the wide range of interests covered by the newspaper will, in turn, assist in developing these math skills.

Advertisements

Use the advertisements in newspapers to help your child become a wise shopper as well as to realize the need for problem solving and accurate computation.

GROCERY ADS: Most newspapers have a weekly section that focuses on local supermarket specials. Use this section to provide your child with meaningful practice of computation skills, "real" problems to solve, and experience with comparison shopping.

Ask your child, as soon as he is able to read 2-digit numbers, to help you estimate the cost of your groceries for the week by reading prices from the paper. Identify only items that cost less than $1. After he learns to read 3- and 4-digit numbers, ask for the price of items that cost more than $1.

As your child develops additional computation skills, ask him for more help in planning your weekly grocery shopping. Suggested activities:

1) Provide a list of items that you need to buy and ask your child to try to find the price of each item at as many stores as he can and determine which is the best buy.
Examples:

Items to buy:	Amount to buy:	Price at Markets: A	B	C	Best buy:
Sliced Bacon	1 lb.	1 lb. $1.79	1 lb. $1.39	1 lb. $1.89	B
Cube Steak	2 lb.	1 lb. $2.99	3 lb. $5.99	2 lb. $2.69	C
Dishwasher Detergent	50 oz.	50 oz. $2.75	50 oz. $2.49	50 oz. $2.56	B

Stress that to compare the cost of two items, the child must:

a) Express the amount of each in the same units (ounces, pounds, feet, yards);
b) divide the total cost of each item by the number of units to find the cost per unit; and
c) compare the cost per unit for the two items.

Ask such questions as, "How much will we save if we buy these items at Market B rather than Market A. How much will we save if we buy them at Market B rather than Market C?"

2) Look for items listed in different measures (units) and discuss comparisons.

Store A
1 qt. Brand X Mayonnaise @ $.99
Store B
32 oz. Brand X Mayonnaise @ $1.29

Ask, "Which is more, 1 quart or 32 ounces?" (Same) "Which store has the better buy?" (Store A)

3) Allow the child to spend a limited amount of money to buy food for a party, a picnic, or any get-together for friends, using the newspaper grocery ads to plan the menu and the grocery shopping list.

4) After your grocery list is made, show the coupons that you have collected and ask, "How much can we save by using these coupons?" If you have not collected coupons, ask the child to look through the paper to find discount coupons for the items on your list and determine the amount of savings.

Discount Coupon

DEPARTMENT STORE ADS: Before your child buys any item of significant cost, encourage the child to shop for value. The newspaper is one source of information that can help the child make "educated" decisions about where, when and what to buy.

Example:

Price of baseball glove: Store A-$29.88 Store B-$32.50

Ask. "Which is less?" ($29.88) "How much less?" ($32.50 - 29.88 = $2.62)

When reading through department store ads, children will see such terms as "20% off" and "30% savings." Though your child probably has not learned to find a percent of a number by the time he has left the primary grades, he may have curiosity about the meaning of terms like these. There is a simple explanation of this process in Chapter 7. Assist with the computation if the child has trouble, and explain how to round the product to the nearest cent. Provide situations for practice similar to those following this explanation of the process.

If a number to the right of the place value to which you are rounding is five or more, round up. If that number is less than five, round down. Examples:

125 rounded to the nearest 10 is 130.

542 rounded to the nearest ten is 540.

Situation:

Store A has a bike listed for the regular price of $89.99 for sale at 20% off. Store B has the same bike for the regular price of $89.88 for sale at a savings of $20. Ask, "Which is the better buy?" Be aware that the child may not have learned to multiply by a 2-digit multiplier, so provide any help that he needs. The child is learning which process to use to solve this problem:

20%=.20

A	B
$89.88	$89.88
x.20	-20.00
$17.9760 = $17.98	$69.88

($17.9760 is approximately equal to $17.98)

$89.88
-17.98
$71.90 (Sale Price)

Since $71.90 is greater than $69.88, Store B has the better buy. Ask, "How much less is the bike at Store B?" ($71.90 - $69.88 = $2.02.)

Remind your child to calculate the amount of sales tax as it applies in your state to find the total cost of any item. This is discussed in Chapter 7.

Example:

5% sales tax (5% = .05)

Bike at Store A	Bike at Store B
\$71.90	\$69.88
x.05	x.05
\$3.595 eqv \$3.60	\$3.4940 eqv \$3.49
(\$71.90+3.60 =\$75.50)	(\$69.88+3.49 =\$73.37)

Difference: \$75.50 - \$73.37 = \$2.13.

After the sales tax is included in each of the above, the difference in price increases to \$2.13.

Automobile Ads:

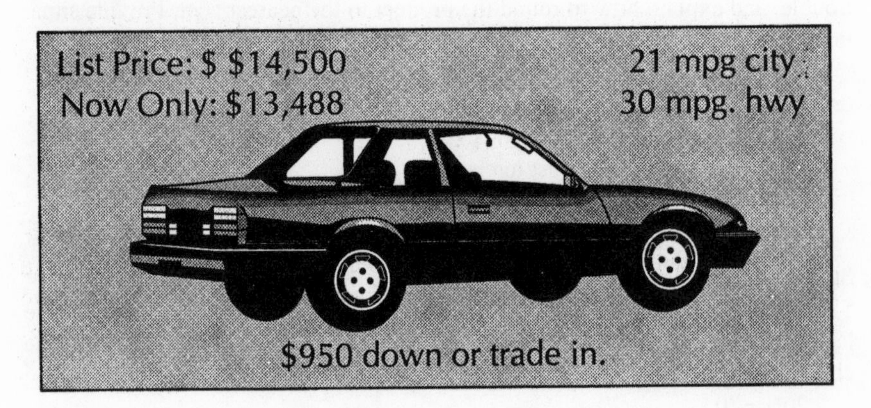

List Price: \$ \$14,500
Now Only: \$13,488
21 mpg city
30 mpg. hwy
\$950 down or trade in.

Many children in third and fourth grades are keenly interested in motor vehicles-automobiles, trucks, RVs, boats and motorcycles. Discuss the newspaper ad for these items with your child by asking such questions as:

"How much has the price of this car been reduced? If we pay \$950 down, how much will we owe? If we trade in our car for \$1,500, how much will we owe?

Find ads in the newspaper that give the estimated gas mileage for a car and ask:

"How many more miles per gallon does this car travel on the highway than in the city? (Discuss reasons: more stops and starts and traffic lights and more idling of the motor in heavy traffic.) If it will travel 21 miles per gallon, how far will it travel on 3 gallons? How much will the gas cost to travel 63 miles?" (Give the price that you pay per gallon.)

"I drive my car about 250 miles each week. How many gallons of gas would I use in this car?" If the child has not learned to divide by 2-digit numbers, ask how you can find this number of gallons. (Divide the number of miles by the miles per gallon.)

CLASSIFIED ADS: Using the classified rate schedule for your newspaper, have your child prepare an ad for publication and determine its cost for 3 days, 5 days or 2 weeks. Encourage the child to study other ads and try to reduce the cost by eliminating unnecessary words or by using abbreviations to shorten the length of sentences or phrases.

Sports Section

Though many third- and fourth- grade students are interested in sports and the records of their favorite teams or athletes, they often do not have the computation skills to calculate many of the statistics. This interest can motivate accurate computation and problem solving. Ask your child simple questions similar to the following, and have the child look for information in the sports section of the newspaper for computing answers.

"Which team scored more runs (or more points) in the game? How many more? Which team has won more games this season (pick two)? How many more?"

In football, for instance, you may ask:

"Which team made more first downs? Rushed for more yards? Passed for more yards? How many more? Joe Brown carried the ball 6 times for 29 yards.

What was his average of yards per carry?

Note: place the remainder (5) over the divisor (6) to form the fraction 5/6

$$\begin{array}{r} 4\tfrac{5}{6} \\ 6\overline{)29} \\ \underline{24} \\ 5 \end{array} \quad \text{or} \quad 29 \div 6 = 4\tfrac{5}{6}$$

Homes Section

Many newspapers have one section a week that focuses on the home-buying or building a home. furnishing a room or home and activities for enriching home life. These offer great possibilities for use of mathematical skills. Though many of the topics are geared to adult interests, motivating factors for your child are. as always, your interest in the child's activities, your patience if the child falters, and your approval of the child's accomplishments. Involve the child in these activities to increase his interests in the home and to reinforce math skills.

RECIPES: Sometimes this section of the newspaper includes interesting recipes. Using these recipes will provide practice for your child in following directions and making accurate measurements. Doubling or halving a recipe will give practice in multiplying fractions, as described in Chapter 6.

Example:

BUTTER COOKIES

2 cups flour
1/2 cup confectioner's sugar
1 tsp. vanilla
1/2 Ib. butter
1/4 cup chopped walnuts

Mix all ingredients, except walnuts, together and form into 1-inch balls or crescent shapes. Roll in nuts and bake at 325 degrees for 10-12 minutes.

To double:

Flour 2x2 cups = 4 cups
Sugar. 2 x 1/2 cups =1 cup
Vanilla: 2 x 1 tsp. = 2 tsp.
Butter: 2 x 1/2 Ib. = 1 Ib.
Walnuts: 2 x '4 cup = 1/2 cup

To halve:

Flour 1/2 x 2 cups = 1 cup
Sugar. 1/2i x 1/2 cup = 1/4 cup
Vanilla: 1/2 x 1 tsp = 1/2 tsp.
Butler 1/2 x 1/2 lb.= 1/4 Ib.
Walnuts:1/2 x 1/4 cup = 1/8 cup

FURNISHINGS: This section often has many furniture ads and articles on furnishing and decorating the home. The material offers opportunities to have your child compare price (Which costs less? How much less?), to find the total cost of many items, and to find the cost of more than one item with the same price. (One lamp costs $89.98. How much will I pay for two lamps?) As with department store ads, there will be discounts and sales taxes to compute. The following activities require these computation skills, measuring skills and present a need to make and read scale drawings.

1) If you are embarking on a redecorating or refurnishing project, ask your child to help with problem solving and to give opinions (likes and dislikes) as often as possible.
2) Ask your child to pretend to be redecorating and refurnishing his bedroom, to choose the furniture and the materials, and to compute the cost. Ask the child to measure the room. list the measures, and make a drawing on graph

paper in order to find out if the furniture will fit into the room. Suggest that one square on the graph paper represents one square foot of the room.

Encourage the child to a) Measure each piece of furniture; b) draw it on graph paper to the same scale (same size squares) as the room; and c) cut out the drawing and place it on the scale of the room in order to determine the best placement. Explain that when you are shopping for furniture, it is helpful to take a scale drawing of the room with you to determine if the furniture will fit the available space.

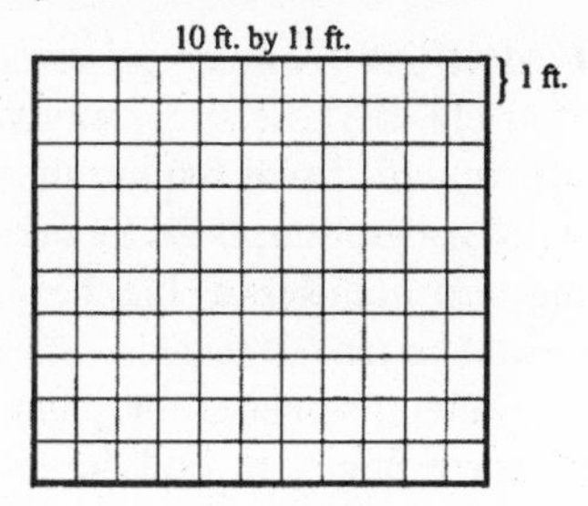

Craft projects: Look for craft projects in the home section of your newspaper. Your child can practice math skills by reading and following directions, and reading scale drawings as he helps you complete some of these projects. Activities that are of interest to many children and are often seen in this section of the newspaper include directions for making birdbaths, birdhouses, doll houses and other toys, storage units, macrame plant hangers and needlepoint pictures.

New Homes: If you are interested in looking at new homes, your child will probably create an interest in looking at house plans in the real-estate section of the newspaper. Sometimes these plans are drawn to scale. Discuss interesting plans, placement of your furniture, and uses made of different space. If you can determine the scale of the house plan, suggest that the child draw some of your furniture to the same scale and see how it would fit into that house. Keep measures to the nearest foot or square foot to simplify the scale.

Ask your child to find an ad for a rectangular building lot, draw it to the same scale as the drawing of a house plan, and place the house on the lot to determine the best placement.

Example: ½ inch represents 10 feet:

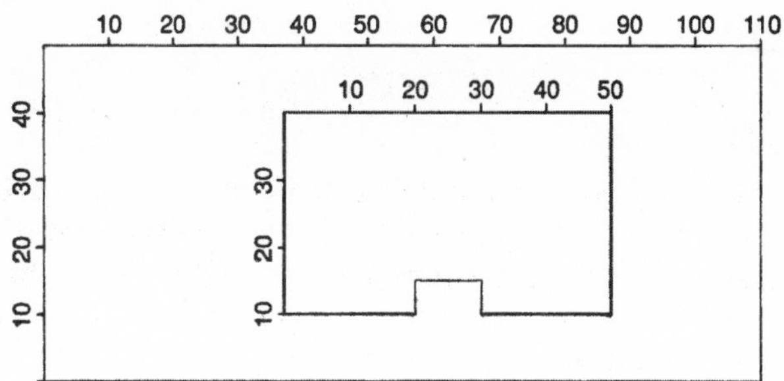

Ads often give the size of house sites in acres and your child may wonder at the size of this measure. The measurement is usually defined in the number of square feet—43,560—but this is difficult for most children to visualize. Explain that a football field, from goal line to goal line, has 48.000 square feet and, therefore, is a little larger than an acre.

GARDENS: Articles on gardening will suggest problem-solving activities for your child. These suggestions include the amount of space needed for various plants, which will require the use of measuring skills. Knowledge of temperatures that are suggested for certain plants will increase the child's interest in reading thermometers-Fahrenheit or Celsius-to help decide when potted plants should be moved inside.

Advertisements for plant and gardening materials offer opportunities to compare prices, to find the total cost of materials, and to solve two-step problems. Examples:

1) The root stimulator is $2.98 per pint and $4.98 per quart. Which costs more.
2 pints or 1 quart? (2 pints) How much more ($0.98 or 980
2) This weed killer will cover an area of 5,000 square feet. The area of our lot is 6.250 square feet, but our house covers 1.300 square feet. Will the weed killer cover the grassy pan of our yard? (6.250 - 1.300 = 4.950. Yes, it will.)

WEATHER: Studying the daily weather summary in the newspaper will often encourage a child's interest in scientific and mathematical problem-solving situations. Instead of reading this section of the paper yourself, ask your third- or fourth-grade child to find information for you. Adjust requests similar to the following to match the information given in your newspaper.

Example:

"What is the weather forecast for today-high and low temperatures? Will it rain? What were the high and low temperatures yesterday? What was the range of temperature (difference between high and low)? How much did it rain yesterday? How much last month? Can you show me on a ruler how much it rained last month?"

Summary:

This chapter offers suggestions for using the newspaper as a practical source of mathematical applications for your child. These activities are designed to increase the child's realization of the importance of mathematics through the use of the newspaper and to increase his appreciation of the news as a source of relevant and current information.

As the two of you become involved with these activities, many other problem-solving situations will arise. The design and content of your daily paper and the mathematical skills of your child will continually change; hence, the possibilities of the newspaper as a learning resource will increase.

9 MATHEMATICS AND TRAVELING

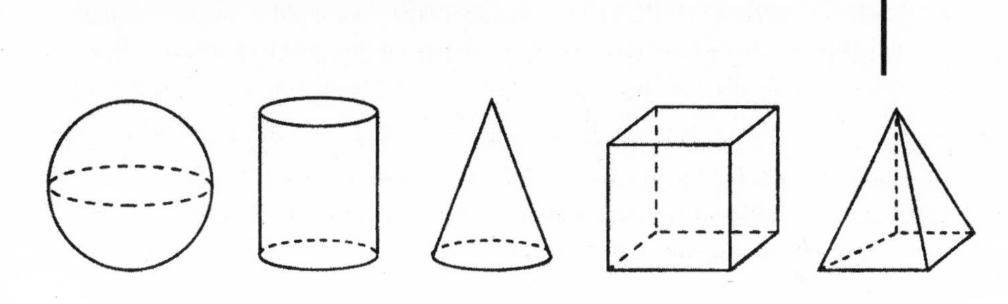

Whether you and your children are taking a thousand-mile trip or making short hops around your neighborhood, many occasions will arise to use mathematical skills while traveling or planning to travel. This is a good time to use your child's natural curiosity to help him discover that mathematics is all around us. The child's interest in cars, planes, trains, or any moving vehicles will add spark to his desire to solve mathematical problems that relate to these means of transportation.

Making Short Hops Around the Neighborhood

FINDING USES OF NUMBERS: As you drive around your neighborhood, help your child observe some of the many places where numbers are used: street signs, addresses, prices in store windows or gas stations, license tags, numbers on houses or businesses. Ask questions that will stress patterns in these numbers.

Examples:

1. As you go one direction on a street, ask the child to write, in a column, the house numbers that are on his right. Change direction and ask him to write, in another column, the house numbers that are now on his right. Ask: "How are the numbers in your first column alike? (They will probably all be even

or all odd.) How are the numbers in your second column alike?" (They will be all odd or all even, the reverse of the first column.) Encourage the child to check other streets to see if the numbers on one side are always even and the numbers on the other side are always odd.

2. Discuss price signs. Ask: "How are all of the money numbers alike? (They all have a dollar sign.) How many digits are to the right of the decimal point?" (Either none ($46.1 or two [$3.56.])

COUNTING AND COMPUTING: To encourage your child to practice counting and estimating and to observe his surroundings more closely, play the following estimating game. Before starting on a short trip, suggest that each person making the trip estimate the number of stop signs (or traffic lights) you will pass on the way. Have each player write his initials and his estimate on a piece of paper and put it in the glove compartment. Each will count the stop signs as you make the trip and upon arrival compare the number counted to the number estimated by each person. The one with the closest estimate wins the game. For each trip. name another item to estimate and count.

Suggestions:

1. The number of yellow cars you will meet;
2. the number of churches (service stations, silos);
3. the number of trucks (vans, boats, campers); and
4. the number of women drivers (men drivers).

Reinforce your child's ability to count and compare (find the difference) by having two passengers (you and the child, or two children) guess whether you will meet more white cars or more blue cars in a specified distance or length of time, and then count these cars as you travel. Suppose, for example, that you see 12 blue cars and 22 white cars. Ask if there are more blue cars or white cars and how many more. After the players have done this several times, have each of them make a generalization about whether more people buy blue cars or white cars. Ask if they can think of ways to check their generalizations. (Visit an automobile dealer to get additional facts on color choices of cars.) Encourage them to make this check.

FINDING SHAPES: On another excursion, look for various shapes. Have your child identify different traffic signs and name the shapes. For example:

stop sign ⟶ octagon
yield sign ⟶ triangle
railroad sign ⟶ circle
information sign ⟶ rectangle

Point out other signs and ask your child to name the shapes.

Solid figures such as cones, spheres, cylinders, pyramids, and cubes may be found in the architectural design of buildings. Show these shapes to your child and discuss their characteristics. A sphere looks like a ball or globe; a cylinder looks like a can or a silo; a cone resembles an ice cream cone or an Indian teepee; a cube looks like a box with square sides or a die (one of a pair of dice);
a pyramid has three or four triangles that come to a point like the roof of a house.

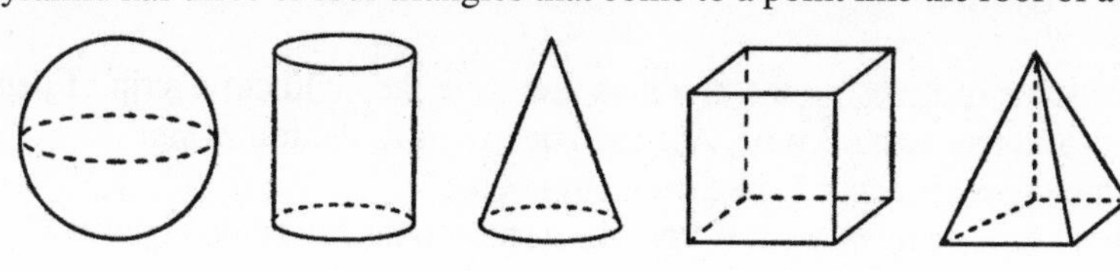

MEASURING DISTANCE: Show your child the odometer on your car and explain that it measures distance in miles (kilometers) and tenths of miles.Watching the movements of the odometer will reinforce the concept that 10 tenths = 1, 10 ones = 1 ten, 10 tens = 1 hundred, and so forth moving to the left. Ask which wheel moves most often (tenths) and why. (A tenth of a unit-mile or kilometer-is the shortest distance measured.)

Have the child record the mileage that is on the odometer before you start for some stated destination, such as the grocery store, the shopping center or Grandma's house. When you arrive, have him record the mileage again and subtract to find the number of miles and tenths of a mile that you traveled.

After the child has figured the actual mileage to several places, have members of the family make estimates of the distance to some specified site. Estimate to a tenth of a mile for a short distance and to a mile for a longer distance. Have him figure the actual distance and the difference between that distance and each person's guess to see whose estimate was the best.

USING CITY MAPS: If you live in a large town or city. you will need to plan many of your trips in advance and this will be an opportunity to help your child learn to read a city map. When you are looking for the location of a street, show how to locate it in the alphabetical street index on the back of the map and note the letter and number that designates the site. Turn to the map and discuss the location of the letters on the sides and the numbers across the top and bottom of me map. Ask the child to locate the street by finding the intersection of the area indicated by the letter and that indicated by the number.

Ask for the letter and number that locates the street on which you live, the school that the child attends, and the city library.

Discuss the legend of the map. Show the direction symbol and ask questions such as the following:

1. "Which direction do we go when I take you to school? When we come home from your school?"

2. "Which direction is to our right if we are going north? (East.) Which direction is to our left when we are going north?" (West.)

3. "Which direction is to our right if we are going west?" (North.)

Show the scale of miles on the city map and have the child cut a strip of paper the length that represents 1 mile. Ask such questions as the following:
1. "Which street is about 1 mile from our house?"
2. "About how many miles is it from the airport to the interstate highway?"
3. "How far is the city park from your school?"

These map reading skills will help him learn to read charts and graphs that are often presented in mathematics.

Planning Longer Trips

When you are planning a long trip, whether by car, airplane, or train, include your child in making some of the preparations.

FINDING AVERAGE DAILY TEMPERATURES AND RAINFALL: Suggest that the average daily temperature for this time of year at your destination will determine the type of clothes you will need to take. Some sources of this information are encyclopedias, newspapers, the Weather Bureau, and the public library. If your TV channels or newspaper report the daily temperature at this city, encourage your child to keep this information for 9 days and then find the average high and average low. Explain that this will tell you what to plan to wear during the day and evening.

CHANGING TIME ZONES: If your trip will take you into other time zones, discuss this with your child. Explain that as we travel across the United States, we cross different time zones. On the east coast, we call the time Eastern Standard Time. Next, moving west, we have Central Standard Time, then Mountain Standard Time, followed by Pacific Standard Time. If we are moving west, we will set our watches back one hour each time that we enter another time zone. (3:00 o'clock will become 2:00 o'clock.) If we are moving east, we will set our watches forward one hour. (3:00 o'clock will become 4:00 o'clock.) Most of these times are changed to Daylight Savings Time in the summer. Find the time zones on a map in the encyclopedia and have the child look at the map and tell you the time in different cities after you have named the time in one zone. Example: "It is 12:00 noon in Washington. D.C. What time is it in Denver, Colorado?" (10:00 A.M.) Before traveling to another time zone, ask him, on several occasions, the time in your own time zone, the time in the place you will be visiting, and what he thinks children in that place are doing at this time.

READING ROAD MAPS: When planning a long auto trip, give your child a road map and ask for help in "mapping out" the route. Show where you are and where you are going. Discuss the symbols in the legend of the map. "Which lines indicate interstate highways? turnpikes? other through highways? How can we tell that a section of a highway is under construction?"

After this discussion, ask the child to choose the route you should take. Discuss whether you want to take the shortest route or to visit places of interest and take the most scenic route.

As he studies the road map, ask: "How are the numbers on north-south interstate highways alike? (They are odd numbers.) How are the numbers on cast-west interstate highways alike? (They are even numbers.) How do north-south interstate highway numbers change from west to east? (The numbers get larger.) How do east-west interstate highway numbers change form north to south?" (They get smaller.) It is not the object of this activity to have your child Learn about highway numbers, but rather to learn to look for and identify patterns in numbers.

Point out the scale of miles and see if the child can estimate the number of miles you will travel to your destination. Show how to mark a strip of paper to match the scale of miles.

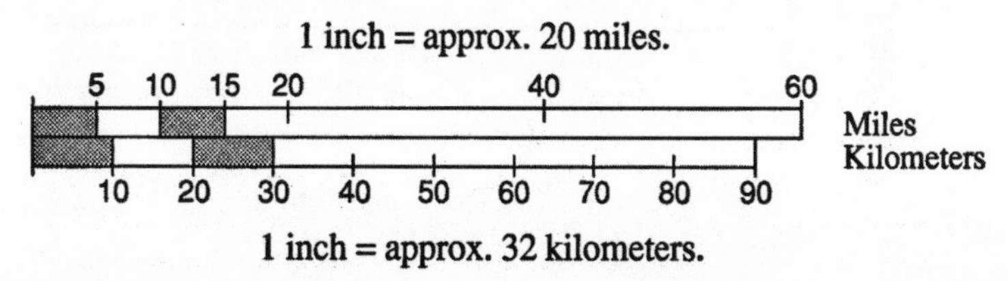

Explain that this is an estimate and that the actual number of miles may be different. Tell the approximate number of miles you will drive in an hour and see if the child can estimate the number of hours of driving time for the trip. Discuss the number of times you will need to stop to eat, the approximate amount of time for each stop and estimated number of "pit stops"-refueling and maintenance, and an approximate amount of time for each stop. Encourage him to estimate the total amount of time for the trip-driving time, plus eating time, plus "pit stops." Save this estimate and compare it with the actual amount of time the trip takes.

READING A TIMETABLE: If your are planning a trip by plane or train. lake that opportunity to help your child learn to read a timetable similar to the one below.

Discuss the meaning of the notations in each column and explain that the colon has been omitted from the times for takeoff and landing. Stress that the two digits on the right tell the number of minutes past the hour. That leaves one or two digits on the left that name the hour. Example: 140a means 1:40 a.m. and 10:24p means 10:24 p.m.

Orlando

TWA

	Leave	Arrive	Flight No.	Stops or via	Freq.	Service
From Los Angeles/ Ontario, Cal When in Los Angeles, call 483-1100	L 140a	107p	436/462	St. Louis	Daily	
To Minneapolis/ St. Paul, Minn	1024a	132p	467	One-stop	Daily	
From Minneapolis/ St. Paul, Minn When in Minneapolis, call 333-6543	810a	107p	462	One-stop	Daily	
To Philadelphia, Pa.	135a	346a	462	NON-STOP	Daily	
From Philadelphia, Pa. When in Philadelphia, call 923-2000	745a	956a	467	NON-STOP	Daily	
To St. Louis, Mo.	1024a	1130a	467	NON-STOP	Daily	
From St. Louis, Mo. When in St. Louis, call 291-7500	1000a	107p	462	NON-STOP	Daily	

Ask questions about the timetable for your flight similar to the following questions about the timetable shown above.

"How long does it take flight number 462 to go from Orlando to Philadelphia, Pa.? (Subtract leaving time from arriving time-2 hrs. 2 min.) How long does it take flight 467 to go from Philadelphia to Orlando? (2 hrs. 11 min.) Do these flights make any slops? (No.) Are the two cities in the same time zone? (Yes.) Do flights 467 or 462 make any slops between St. Louis and Orlando? (No.) Compare the time that it lakes to fly from Orlando to St. Louis (1024a toll30a-1 hr. 6 min.) to the lime that it lakes to fly from St. Louis to Orlando (lOOOa to 107p-3 hrs. 7 min.) Why is there so much difference? (They are in different time zones.) What is the actual flying time both directions?" (2 hrs. and 6 or 7 min. To find the actual time, convert all times to a common time zone and subtract leaving time from arrival time for both flights.)

ESTIMATING EXPENSES: As you plan your trip. discuss your estimated expenses with your child and ask for help with the computation. As always, if he has not learned the necessary computation skills, just ask the child to tell you what process to use.

Examples:

1. After the child estimates the number of miles you will be driving, tell the number of miles per gallon that your car travels and ask how you can find the number of gallons you will need. (Divide the number of miles for the trip by the mpg.) After you find the number of gallons the trip will take. tell the estimated price per gallon and ask how you can find the approximate total cost of gasoline for the trip. (Multiply the cost per gallon by the number of gallons.)
2. Tell the amount that you anticipate spending for lodging each night and the number of nights you will be paying this amount. Ask how you can find the approximate amount thai you will spend for lodging.
3. Tell the amount you expect to spend for breakfast, lunch, and dinner each day and ask the child to find the total amount food will cost each day and the total amount for the entire trip.
4. Follow the same procedure with other expected expenses and ask him to estimate the total expenses of (he trip.
5. If you are flying or traveling by train or bus, give the cost of the round trip tickets, the cost of food and lodging and miscellaneous expenses and ask the child to estimate the total cost of the trip.
6. Discuss the child's possible expenses-cost of recreation, souvenirs, gifts, etc.-and ask for an estimate of his personal expenses for the trip.

Save these estimates and then compare mem with actual costs after the trip is completed.

Making Long Trips by Car

Opportunities arise on long auto trips for children to practice mathematical skills. Keeping them busy with some of the following suggested activities will make the trip more interesting and long rides in the car less tiresome.

USING THE ODOMETER: Before starting on a long auto trip, provide a pad of paper and a pencil and ask your child to record the odometer reading and the time in hours and minutes-for example, 7:35 a.m. Each time you stop for gas, have him record the odometer reading and the time in hours and minutes. Then ask bow many miles you have driven since the last stop and how much time has elapsed. By writing the odometer readings and subtracting, the child will probably be able to tell how many miles you have driven, but it is a little more complicated to find the amount of time that has elapsed. Remember, if it is necessary to regroup when subtracting a number of minutes, 1 hour must be renamed as 60 minutes. **For example:**

$$\begin{array}{r} 10{:}15 \\ \underline{-7{:}35} \end{array} \qquad \begin{array}{r} \scriptstyle 9\ \ 75 \\ \not{1}\not{0}{:}\not{1}\not{5} \\ \underline{-\ 7{:}35} \\ 2{:}40 \end{array} \qquad \begin{array}{l} (60 \text{ min} + 15 \text{ min.} = 75) \\ \\ (2 \text{ hrs. } 40 \text{ min.}) \end{array}$$

The child may also have trouble if the hour hand passed twelve during the elapsed time.

Example:

Beginning time: 9:35 a.m.
Ending time: 2:45 p.m.

Method A.

(Think 12 + 2 = 14)

$$\begin{array}{lr} \text{Ending time.} & \overset{14}{\not{2}}{:}45 \text{ p.m.} \\ \text{Beginning time.} & \underline{9{:}35} \text{ a.m.} \\ & 5{:}10 \quad (5 \text{ hrs. } 10 \text{ min.}) \end{array}$$

Method B.

Step 1. Find the difference between the beginning time and 12:00

$$\begin{array}{r} \scriptstyle 11\ \ 60 \\ \not{1}\not{2}{:}\not{0}\not{0} \\ \underline{-9{:}35} \\ 2{:}25 \end{array}$$

Step 2. Find the total of that difference (2:25) and the length of time between 12:00 and the ending time. (2:45)

$$\begin{array}{r} 2{:}25 \\ \underline{-2{:}45} \\ 4{:}70 \end{array}$$

Step 2. 70 min. is 1 hr. 10 min. Regroup to hours and write the total hours and minutes. (5 hr. 10 min.)

After the child has found the distance that you have driven in a certain amount of time, ask how to find the average number of miles you have driven each hour. (Divide the number of miles by the number of hours.) Round the hours and minutes to whole hours (2:40 is approximately 3 hrs.) and keep the number of hours below ten unless the child has learned to divide by 2-digit numbers. If he has found the average number of miles per hour that you have driven each day for several days, discuss why you drove faster on some days than others. (Less traffic, split highways, interstate highways, etc.)

Ask your child to keep a record of the number of gallons of gasoline it takes to fill the tank of your car and the number of miles the car travels between fill-ups. Ask how to use these numbers to find the number of miles your car travels on 1 gallon-mpg. If he has learned to divide by a 2-digit number, let him find the mpg. Otherwise, you do the dividing. Compare the mpg that you got with that listed in ads for your car.

USING ROAD MAPS AND MILEAGE CHARTS: Have your child look at a road map and predict which city you will be nearest in one hour (two hours, four hours).

Using the formula $d = r \times t$ (distance equals rate times time) will help in this prediction. For example: If you are traveling an average of 50 mph (rate), in 2 hours (time) you should be 100 miles (distance) down the road. (Distance = 50 mph x 2 hrs.)

Look at the map and the scale of miles to see which city is approximately 100 miles away from your present position. If you do not wish to use the scale of miles, show the child how to read, and then add, the little numbers between cities.

When you are in a city that is listed on the mileage chart of your map, ask your child to use the chart to find the distance to another city that is listed on the chart. Show him how to find the intersection of the column of numbers from one city and the row of numbers from the other city and explain that the number in this space tells the distance between the two. Shown here are two types of mileage charts.

	Albemarle	Ashboro	Asheville	Blowing Rock	Boone	Burlington	Cape Hatteras
Ashboro	41						
Asheville	157	181					
Blowing Rock	117	132	86				
Boone	128	124	95	9			
Burlington	79	39	200	149	140		
Cape Hatteras	350	308	487	443	431	297	
Chapel Hill	95	54	224	174	165	26	271

Asheville to Burlington — 200 miles

	Amarillo	Atlanta	Baltimore	Birmingham	Bismarck	Boston
Albuquerque, NM.	284	1381	1829	1251	1088	2172
Amarillo, TX		1097	1554	967	948	1897
Atlanta, GA.	1097		645	150	1495	1037
Baltimore, MD.→	1554	645		773	1499	392
Birmingham, AL.	967	150	773		1433	1165
Bismarck, ND.	948	1495	1499	1433		1794

Baltimore to Bismarck — 1,499 miles

1. Find the distance between Asheville and Burlington.
2. Find the distance between Baltimore, MD. and Bismarck, ND.

Recording Expenses: Encourage your child to keep a record of his own expenses on the trip — gifts, snacks, recreation, souvenirs — analyze how the money was spent, and compare the actual expenditures with the estimates that he made before the trip began. This activity may help the child spend money more wisely on the next trip. If you are keeping a record of your expenses and the categories of your expenditures, allow him to see your method and determine whether or not to use the same plan. Even if he wishes to assist with your recording, do not let the responsibility become a burden.

CHANGING TIME ZONES: Show the lines on the map that indicate time-zone borders and ask your child ta be responsible for telling you when you should change the time on your watch. If you make any long distance phone calls to your home, discuss the difference in time and whether or not it is an appropriate time to call. Example: "It is 8:30 p. m. here. What time is it at home? If it is 10:30 there, do you think our friends might be asleep?"

CHANGING ALTITUDES: As you travel into, or out of, the mountains, discuss the changing altitude. Explain this term, usually expressed in feet, as the distance above sea level. Encourage the child to find the altitude of places that you visit or spend the night by asking people who live and work in the area. Then ask the child: "Is the altitude more or less than that at home? How much more or less?"

FINDING AND COMPARING AVERAGE DAILY TEMPERATURES: As you travel, encourage your child to check average high and low daily temperatures in places that you visit or spend the night Discuss these temperatures, and how and why they change.

Examples:

1. "Is it colder or hotter high in the mountains than at lower levels?" Explain that if all other conditions are the same, including the distance from the equator, average temperatures decrease a little more than 3 degrees Fahrenheit for each 1,000 feel increase in altitude. Mention that Denver, Colorado is nicknamed the Mile-High City, and ask:
   "About how many feel above sea level is Denver? (5,000) About how many degrees cooler will Denver be than another city at sea level if all other con ditions are the same?" (15)
2. If you are traveling great distances north or south, ask: "How are the aver age temperatures changing as we travel north (south)? Why?" (As we get farther from the equator-north-average temperatures are cooler. As we get nearer the equator, average temperatures are warmer.)

JUST FOR FUN: There are many activities that you can do that involve looking at license plates. The following activities are fun, but they will also help your child learn the rules of divisibility. For the greatest effect, use one at a time, possibly devoting each day to a different activity.

Examples:

1. Give a point to the first passenger that determines whether the license number of the car you are following is evenly divisible by 2 (no remainder when divided by 2).
2. Give a point to the passenger who first identifies a license number that is evenly divisible by 3. A quick way to check each number is to add the digits and see if the sum is divisible by 3. If so, the number is divisible by 3.

$$15 \rightarrow 1+5=6 \rightarrow 6 \div 3 = 2$$

Example:

ALT – 591

$5 + 9 + 1 = 15$

$15 \div 3 = 5$

Check: $3\overline{)591}$ = 197
```
  197
3)591
  3
  29
  27
   21
   21
```

15 is divisible by 3 so 591 is divisible by 3.

If necessary, the divisibility of 15 by 3 can be checked by finding the sum of the digits and dividing by 3.

3. Make a game of finding license numbers that are evenly divisible by 4. Hint: If the last two digits on the right are divisible by 4, the number is divisible by 4.

Example:

JUT – 576 ($76 \div 4 = 19$)

```
       144
Check: 4)576
         4
         17
         16
          16
          16
```
(Yes, because $76 \div 4$ has no remainder.)

JUT – 577 ($77 \div 4 = 19$ R. 1)

```
  144
4)577
  4
  17
  16
   17
   16
    1
```
(No, because $77 \div 4$ has a remainder.)

4. Let finding license numbers divisible by 5 be the game of the day. Hint: If the digit on the right—in the ones place— is 0 or 5, the number is divisible by 5.
5. This one will require two steps. Look for license numbers that are divisible by 6. The number must be divisible by 2 (even) and by 3 (The sum of the digits is divisible by 3).

Examples:

STR – 46 — is divisible by 2, but the sum of the digits (10) is not divisible by 3; this implies that 46 is not divisible by 3 and , therefore, is not divisible by 6.

FUN – 399 — The sum of the digits (21) is divisible by 3, but the number is not divisible by 2; therefore the number is not divisible by 6.

42 – 1236 — The number is divisible by 2, and the sum of the digits is divisible by 3; therefore the number is divisible by 6.

6. This one resembles activity number 2. Look for license numbers that are divisible by 9. If the sum of the digits is divisible by 9, the number is divisible by 9

Example:

| 9 2 4 6 8 5 2 |

9 + 2 + 4 + 6 + 8 + 5 + 2 = 36

36 is divisible by 9, so 9,246,852 is divisible by 9.

Traveling by Plane, Train, or Bus

These means of transportation often add interest to a trip, and this interest, for the elementary school child, includes the use of numbers. Encourage your child to investigate these numbers and compare them to familiar aspects of his environment.

When traveling by jet, discuss the altitude at which the plane is flying-possibly 30,000 feet. Ask: "One mile is how many feet? (5,280) We are about how many miles high (6)? Can you think of any place that is about 6 miles from our home? If not, when we go home we can take a six-mile drive and find out about how high we flew. Can you find how high the astronauts travel when they orbit the earth?"

Discuss the average speed of any means of transportation and compare it with the average speed of your car. Use the formula discussed earlier in this chapter, distance = rate x lime. to find how far you will travel in a certain number of hours-2, 5, or 9 hours. If you are traveling by plane, find how long it will take to reach the first stop and ask your child to find the approximate flying distance. If he has not learned to multiply by fractions, round the time to whole hours before he multiplies.

Examples: 1 hr. 15 min. = 1 1/4 hrs. is approximately equal to (=) 1 hr. 1
1 hr. 50 min. = 1 5/6 hrs = 2 hrs.

Ask such questions as the following about other means of transportation. "Do you know the speed at which the astronauts travel?" (Approximately 16,000 to 17,000 mph.) "Do you think the bus (or plane) travels more or less miles on one gallon of fuel than your car? (Less.) Why?" (The engine with more horsepower, stronger engine, will require more fuel.)

Encourage your child to get as much information about the means of transportation you are using as possible. A steward or stewardess will probably enjoy answering the youngster's questions about the plane, and there is usually information in the pocket in front of each seat. The driver of the bus or the conductor on the train can provide interesting information as you travel with them.

Summary

The trip is over and the suitcases arc unpacked, but there arc still opportunities for your child to use mathematical skills. The following activities will provide practice for these skills.

1. If the child used a road map to estimate distances before the trip and the odometer to measure actual distances while traveling, now he can compare these estimates with actual distances by subtracting to find the differences.
2. If the child estimated expenses before the trip and recorded expenses while traveling, he can compare this estimate to the actual expenses by subtract ing.
3. If you traveled by plane, encourage the child to locate a landmark as far from your house as the highest altitude of your plane. Measure the time it takes to travel this distance.
4. If the child used the formula d = r x t to estimate distances, he can use the same formula to estimate shorter distances and use the odometer to check these estimates.
5. If the child used road maps and mileage charts to find distances to cities on your trip, he can use them to find distances to other cities and interesting locations.
6. The child can continue to seek information about planes, trains, buses, and other means of transportation. Examples: How fast does the Concorde fly? a ship travel? a monorail travel?
7. If the child has compared average daily temperatures of one area with a noth er and discussed the reasons for the differences, he can discuss the difference in temperatures of various cities as they are shown daily on TV. Encourage the child to find the hottest spot in the nation and the coldest spot for the same day, locate the two places on a map, and try to determine the reason for the great difference.

10
MATH POTPOURRI

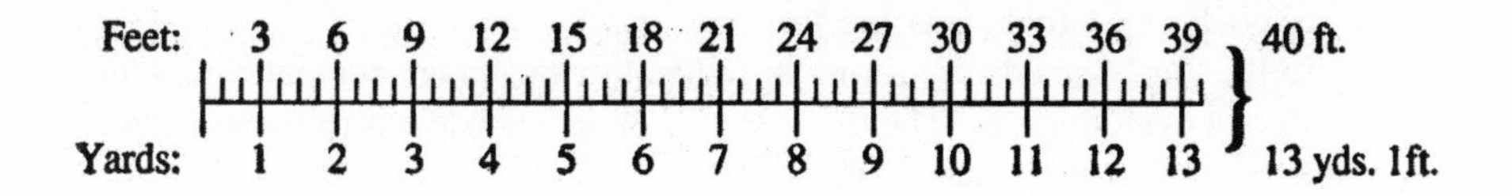

This chapter deals with the mathematical problem-solving possibilities and needs of many areas in a child's life. We hope these suggestions will lead to other activities that will encourage your child to apply mathematical skills and concepts.

Sports

Your child's interest in sports, as a participant or as a spectator, provides great opportunities for interesting and challenging mathematical experiences. These experiences include computation development and practice, measuring activities, and scale drawings.

GAME SCORING: Use the games in which your child exhibits the greatest interest to encourage the use of computation skills.

Examples:

1. Keeping score for your bowling game will give lots of addition practice.
2. Keeping score for a basketball game or discussing the game will require addition, multiplication and division by 2. Include remarks and questions similar to the following in your discussion:
   a. "You shot four field goals. How many points did you score?" (4x2=8)
   b. "Sue scored 10 points with field goals. How many goals did she make?"

$$2\overline{)10}^{\,5} \text{ or } 10 \div 2 = 5$$

c. "David made three field goals and four foul shots. How many points did he score?" (3x2) + (4xl) =
6 + 4 = 10

If necessary, remind the child to solve the little problems inside paren theses first.

d. "Mary scored 14 points in her first basketball game. Name some ways she could do this." (5 field goals and 4 foul shots; 7 field goals: 4 field goals and 6 foul shots, etc.)

3. Use the four methods of scoring in football to encourage computation with different operations. a. "The Bears scored three touchdowns (6 points each), one point after the touchdown, two field goals (3 points each), and a safety (2 points). What was their score?" (3 x 6) + 1 + (2 x 3) + 2 =
18 + 1 + 6 + 2 = 27

b. "The Lions scored 19 points. Who scored more points, the Lions or the Bears? How many more?" (The Bears scored 8 more.) C. "The Mountaineers scored 16 points in their game. Name some ways they could make this score." (1 touchdown, 1 point after the touchdown, and 3 field goals; 2 touchdowns, 2 points after the touchdowns, and 1 safety. Answers will vary.)

4. Many children like to discuss their baseball or softball games. Ask questions about your child's team, or favorite team, that require some computation.
   a. "What was the score of the game? How many more runs did your team make than the other team?"
   b. "How many runs did you make today? How many have you made this season before today?"

MEASURING: Your child's interest in sports can provide experience with measures that will help develop skill in estimating distances, a skill that is important in many endeavors.

After the child has learned to measure with feet and yards and knows the relationship between the two (3 ft. = 1 yd.), discuss the size of the playing field of a favorite sport. To reinforce the meaning of these measures, suggest that the two of you walk each specified distance, count your steps, and compare the two numbers.

Example: Length of a football field, goal line to goal line: 100 yards. Ask: "How many yards long is the football field? (100) How many steps long? Your steps? My steps? Why are there more of your steps than mine? (There are more of the smaller steps.) How many feet long is the football field? (300) Why are there more feet than yards?" (Feet are smaller. Feet and yards are in a 3 to 1 ratio. (3 ft. = 1 yd.)

Use activities similar to the previous example with the following distances:

1. Distance between horseshoe peg boxes (usually 40 ft.). After the child has found the number of feet, ask the number of yards between the boxes. (40 ÷ 3 = 13 yards, 1 ft.) Use a number line sketch if necessary.

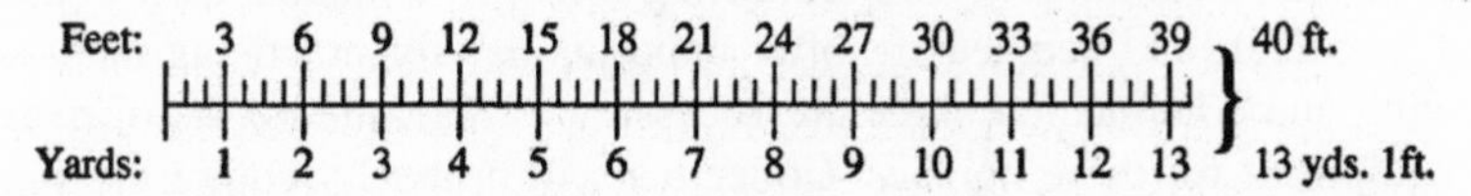

2. Length of one side of, or distance around, a baseball or softball diamond.
3. Length, width, or distance around a soccer field.

Encourage the child to use familiar distances to estimate other distances.

Example:

Say: "How long is one side of our yard? Is it as long as a football field? Is it more or less than 100 yards? Is it half as long as a football field? How many yards is that? Measure and check your estimate."

The sport of fishing also motivates measuring skills. At the end of a fishing trip, encourage your child to measure the length and weight of his catch and compare these measures with those of fish caught by someone else.

SCALE DRAWINGS: The shapes of the playing area of various sports lend themselves to simple scale drawings. (See Chapter 6.)

Example: Football field, 100 yds. from goal line to goal line and 160 ft. (53⅓ yds.) wide.

Scale: ¼ inch = 5 yards

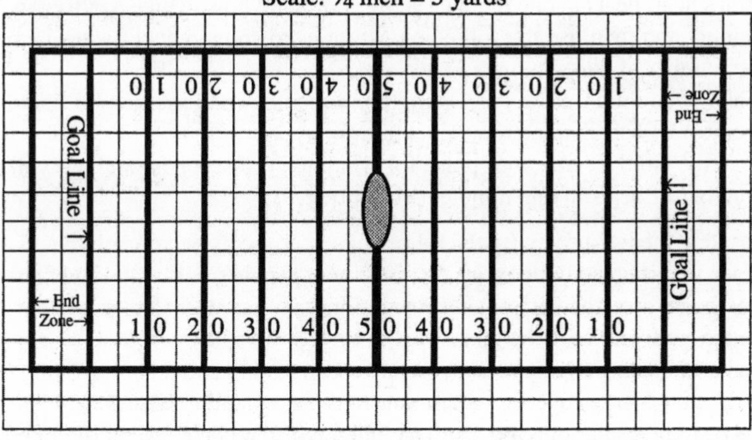

After your child has made the drawing above, suggest that the two of you play Facts Football as described in Chapter 11. Drawing this game board to the scale of *1 inch = 10 yards* and comparing it to the drawing above will help him see the relationship of the same shape drawn to different scales.

Hobbies

Children's hobbies, ranging from collecting (seashells, stamps, etc.) to complicated woodcraft and needlecraft, offer opportunities for practicing math skills. Counting and computation skills are necessary to determine costs of materials and equipment for these hobbies. Collecting will probably require the skills of counting and sorting by likenesses and differences (Chapter 1.) Such crafts as making models, woodwork, and needlework involve the skills of reading and following directions, reading and interpreting diagrams, and measuring to determine the amount of materials needed, size of the finished product, and placement of parts. Use your child's interest in a hobby to motivate practice of any math skill that applies.

Personal Math

Children are usually interested in learning facts about themselves, how these facts change, and how they compare to other people. Finding and using personal numerical facts will provide practice for measurement, computation, and problem solving.

TAKING A PULSE RATE: Explain that a pulse rate tells the number of times a person's heart beats in one minute. Show your child how to take a pulse rate for one minute and then use the following activities to show quicker ways of finding the number of beats in one minute.

1. Take your child's pulse forV! minute (30 seconds) and ask how to find the number of beats in one minute. (Multiply by 2)
2. Ask the child to take your pulse for '4 minute (15 seconds) and tell you the number of beats in one minute. (Multiply by 4)

Suggest that the two of you try the following experiments on yourselves and others to see if certain activities have an effect on a person's pulse rate. Explain that an experiment must be performed several times and on more than one subject to assure the results.

1. Take each other's pulse when you have been resting and record the rate. Run in place or jump rope for one minute and then take the pulses again and record the results. Ask the child to find the difference between (he first and second rates.
2. Take each other's pulse before and after eating and record the results. Ask if there is a difference and, if so, how much.

3. Take the pulses before and after showering. Ask if there is a difference and, if so, how much.
4. Encourage the child to suggest activities that might change the pulse rate and to check them out.

Taking Body Measurements: Measuring and comparing heights and weights will help your child learn to make accurate measurements and provide practice for subtraction. Use standard English units (inches, feet; pounds) or metric units (centimeters, meters; kilograms).

It will be simpler to compare heights given in inches as in exercise "a" than those in feet and inches as shown in exercise "b."

1 ft. = 12 in.

	a.	b.	
Joe's height:	74 inches	6 ft. 2 in. =	5 ft. 14 in.
Mary's height:	54 inches	−4 ft. 6 in. =	4 ft. 6 in.
Difference:	20 inches		1 ft. 8 in.

You can buy a meter tape or help your child make one on ribbon or on paper by cutting out strips, gluing them end to end and marking centimeters from this picture of a ruler. Make the tape 1 meter (100 centimeters) long.

0 1 2 3 4 5 6 7 8 9 10

Measuring various parts of the body with a metric tape will help your child become familiar with the metric system, estimate distances in metric measures, and see relationships between the measures. Encourage your child to find the following measures and record them on a chart similar to the one below.

After the child has measured and recorded the measures, ask questions about their relationships.

Examples:

1. "Each person's foot is about the same length as what other part of the body?" (Forearm)
2. "Is the calf of the leg about the same size as any other part of the body?" (Probably the neck)

Measures in centimeters:	Kim	Bob	Sue
Thumb length:			
Foot length:			
Forearm length: (elbow to wrist)			
Arm length: (Shoulder to wrist)			
Around thumb:			
Around wrist:			
Around neck:			
Around head:			
Around calf of leg:			
Height:			

3. "How does the distance around the thumb compare to the length of the thumb?" (About the same.)
4. "The distance around the wrist is about how many times the distance around the thumb?" (2 or 3)

In order to estimate distances, a child must be able to estimate the unit length. Help your child find a reference point for the size of a centimeter by giving the following directions: "Measure the width of each of your fingers and see if you can find one that is about one centimeter wide. Use this finger width to estimate the length of a book (pencil, telephone, or any handy object) in centimeters. Now, use the meter tape to check your estimate."

To give more practice in estimation, make a game of having a leader name a distance (the length or width of a sheet of paper, the height of a doll) and have two people estimate the distance in centimeters. Give a point to the person that made the closest estimate. Continue until one player gets 5 points.

Use the following activities to give your child a reference point for the size of a meter:

1. Ask: "Are you more or less than 1 meter tall? How many centimeters more. or less? Let's see where the tape reaches on you if we hold it straight up from the floor. About how many meters wide is this room? the hall? How high is the ceiling of the room? Measure with the meter tape to check your estimates."
2. Make a game of estimating distances in meters as you did with centimeters, but don't expect him to estimate, with much accuracy, distances greater than those listed above.
3. Ask: "How many centimeters do you think you can jump (step. hop) from this line? Try it and I will mark the place where you land." After the jump, say: "Did you jump more or less than one meter? Estimate your jump in meters or centimeters. Now measure the distance to check your estimate."
4. Let two people jump from the line, estimate the distance of each jump and the number of centimeters between the length of their jumps. Then have them measure the distances to check their estimates.
5. Follow the same procedures as in activities 3 and 4 to have the child estimate and find the number of meters that he can throw a ball or a frisbee.

MEASURING TIME: Use questions similar to the following for discussion of how you and your child spend most days: "Do we make wise use of our time? Do we spend too much time on one activity that could be spent to better advantage on another?"

A chart that includes your major activities for 5 days (Monday a.m. to Saturday a.m.) will give a good picture of what each of you do each day. Find the average number of hours you spend with each activity and how many hours are left for miscellaneous involvements.

Example of a child's chart: Each day includes 24 hours. Monday runs from the time you get out of bed on Monday until you get out of bed on Tuesday. Friday extends to the time you get out of bed on Saturday.

Monica Becker's Daily Chart	Hours Each Day:					
	Mon.	Tues.	Wed.	Thurs.	Fri.	Avg.
Sleeping	9	10	8	9	10	$9\frac{1}{5}$
School and traveling	7	7	7	7	7	7
Doing chores	1	1	1	1	1	1
Doing homework	1	1	2	2	0	$1\frac{1}{5}$
Playing	1	2	2	1	2	$1\frac{3}{5}$
Watching T.V.	1	1	1	1	2	$1\frac{1}{5}$
Total for each day	20	22	21	21	22	$21\frac{1}{5}$

Ask: "What is the average number of hours not accounted for each day? (24 - 21 1/5 = 2 4/5) What do you do during that time? (Eat, read, rest, talk with parents and brothers and sisters, take music lessons, etc.) Do you think that you should devote more (or less) time to any activity?"

Fun with Math

Just as many children read for recreation, so do many find the patterns and relationships of numbers amusing and entertaining. Just as recreational reading improves word attack and comprehension skills, recreational math improves computation and problem-solving skills. Throughout this book we have included games, puzzles, tricks and other math activities that appeal to a child's interest in anything presented with an air of mystery, but the supply of these items is great. Your child will probably learn some in school and want to share them with you. Be eager to learn; enjoy the tricks and show an interest in the fascinating patterns.

STRATEGY GAMES: Some of these recreational activities will help your child learn to plan strategy, which will lead to better problem-solving skills. The following are examples of these strategy games:

1. Nim is one of the oldest strategy games. There are many variations, and here is one example. Place 20 toothpicks in a row. Each of two players will alternately pick up one, two, or three toothpicks until they are all picked up. The player who picks up the last toothpick loses the game. At first, players will usually decide randomly whether to pick up one, two, or three toothpicks, but as they play the game over and over they begin to form a strategy and become more adept at the game.

For other variations of the game, change the limit on the number of tooth picks that each player is allowed to pick up. Ask: "Does the strategy change when you can pick up more sticks? Why or why not?"

2. Traditional Tic Tac Toe is also good for helping a child plan effective strategy. There are many variations of this game. Coordinate Five in a Row is a version that will help develop strategy and help the child learn to use coordinates (pairs of numbers) to locate points on a graph. Prepare a gameboard by marking the coordinates on a sheet of graph paper. Two players take turns writing an ordered pair of numbers and placing X or 0 at the point indicated by the pair. The first player to get five marks in a row, either horizontally, vertically, or diagonally, is the winner. To use ordered pairs to locate points, it is extremely important to follow the rule of "over, then up." Example: 3, 5 means to start at 0, move three places to the right, and then five places up. Here is a sample gameboard with the first seven ordered pairs of two players and their locations on the graph.

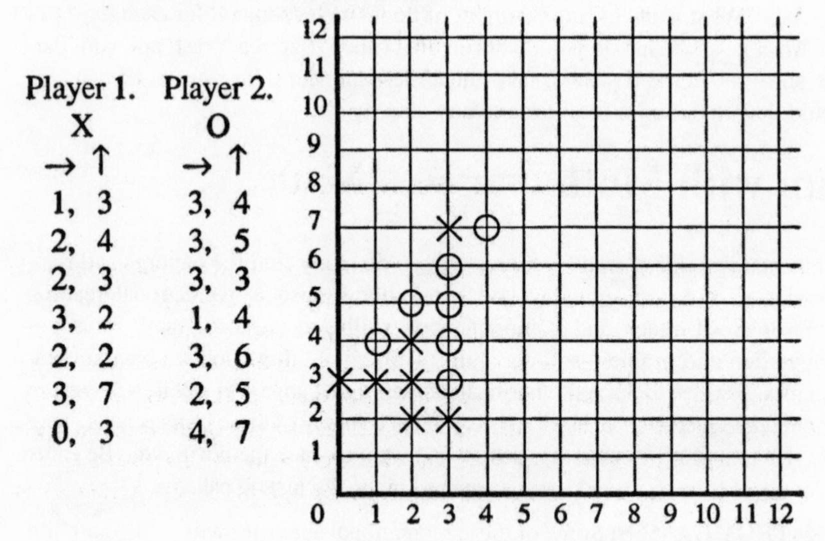

Kalah is an excellent commercial strategy game that is usually enjoyed both by adults and children. If this game is available, play it often with your children and watch their strategies improve.

TRICKS AND PUZZLES: Some of these activities require the use of paper and pencil, but some can be played while riding in a car or doing household chores. Approach these with an air of mystery and lake turns being a detective to find the solution

1. Name the Numbers. Say that you are thinking of two numbers, that you will give two clues and ask the child to find the numbers with those clues. a. "The sum of the numbers is twelve and the difference between the numb-.-

bers is two. Can you name the numbers?" (7, 5: 7 + 5 = 12 and 7 - 5 = 2) b. The sum of the numbers is twelve and the product of the numbers is 32. Can you name the numbers? (4, 8: 4 + 8 = 12 and 4 x 8 = 32) Ask the child to think of two secret numbers, give you two clues similar to those above, and let you try to name the numbers.

2. Sequences are challenging. Begin with easy patterns and gradually increase the difficulty. Show sequences similar to the following and ask the child to continue with the same pattern. a. 1.3,5.7,9, _,_,_(! 1,13,15; odd numbers) b. 1.2,4, 8, 16, _, _, _ (32,64,128; multiply each preceding number by two) c. 1, 1, 2, 3,5,8, _, _. _ (13.21, 34; add the two preceding numbers) Encourage the child to prepare sequences for you to continue.
3. Finding a Birth Date. Tell the child to write down the number of his birth month. (January is 1 because it is the first month of the year; February is 2 because it is the second month of the year; March is 3 because it is the third month of the year, etc.) For example, let's assume his birth month is June, so he will write 6. Now tell him to multiply this number by 2 (2 x 6 = 12); add 5 (12 + 5 = 17); multiply by 5 (17 x 5 = 85); multiply by 2 (85 x 2 = 170); multiply by 5 again (170 x 5 = 850) add the date of his birthday (If his birthday is June 28, he will add 28. 850 + 28 = 878; subtract 250 (878 - 250 = 628). Ask the child to find his birthday in the answer. The number of the month should be on the left and the number of the day of the month should be on the right. (6/28)
Encourage your child to practice giving these directions to family members and then try it on friends. When a friend tells the answer, your child can name the other child's birthday-month and date.
4. Make Fifteen. This is a game for two players and will provide practice for building and solving equations (number sentences). In turn, each player throws three dice and uses the number showing on the top of each die to form equations naming numbers 1 to 15, in that order. The number on each die must be used once, and only once, in an equation. When a player is unable to form an equation that names the next number, the play passes to the other player.
Example:
The first player throws 2,4, and 6 and forms the equations on the right. This player is unable to name 6, so the dice are passed to the second player who follows the same procedure. Player number one will begin naming numbers at six on the next round of play. The first player to name all of the numbers to 15 is the winner; however, if both players reach 15 in the same round, extend the goal to 21.

$(2+4)\div 6=1$
$(2 \times 4)-6=2$
$(2 \times 6)\div 4=3$
$(6+2)-4=4$
$(6+4)\div 2=5$

Calculators

Computers and calculators will be a pan of children's lives today and in the future, both in and out of school. The availability of hand-held calculators makes it possible for you to help your children become comfortable with these tools during their early years. You can use the calculator to reinforce your child's math skills, relieve the boredom of tedious computational tasks and increase interest in learning math.

Your child must know basic facts and have computation skills and problem-solving techniques in order to use the calculator efficiently. These skills are necessary to know which process to use to solve a problem and to be able to check answers by mental estimation and approximation. It is also necessary to know that a certain thing must be done in a certain way.

Examples:

a. To record 3 lens and 5 ones (35), press 3 and then 5. To record 5 tens and 3 ones (53), reverse the order.

b. To solve 18 +3, press keys in the order: 18 + 3 ".

Activities similar to the following will help improve your child's math skills and ability to use the calculator and emphasize (he importance of estimating and approximating accurately.

1. When your child is learning to record numbers, show sets of objects and ask him or her to show the number of each set on the calculator. Examples:
   a. 6 buttons (6);
   b. 2 sets of ten and 5 toothpicks (25); and c. 3 dollars and 65 cents (3.65).
2. Name numbers orally or write number words and ask the child to show the number on the calculator.
   a. Say: "five hundred six" (506);
   b. write: six hundred five. (605); and c. say: "two hundred fifty" (250).

3 If your calculator has a constant function key. play a calculator version of What's My Rule. The leader keys in the same secret number and process before each player enters a number. Two players take turns entering a numb-
\- ber and trying to guess the rule.

Example:

The leader keys in 4 +. The first player enters 6 and gets 10 as a result. The second player enters 5 and gets 9 as a result. The first player enters 7, gets 11 as a result, names +4 as the rule, and wins the game.

4. In mathematics, the accepted order of operations is as follows:
   a. Solve the process inside parentheses;
   b. multiply and divide; and

c. add and subtract.

Key in 3 + 4 x 2 = . The calculator will follow the above order by multiplying, 4 x 2 = 8, and then adding, 8 + 3 = 11. You can change the order. Press 3 + 4 and then =. Follow this by pressing x 2 and most calculators will multiply the sum of 3 + 4 by 2 and give the result of 14. If your calculator has parentheses, press the appropriate one of these keys before and after the part of the problem that you want to solve first.

To provide practice for changing the order of operations by placing parentheses in an equation, play The Greatest Number. Write a number sentence with three 1-digit numbers and +, –, x, or ÷ signs. Have each player find an answer and write it. After all players have found a solution on the calculator, compare the answers. The player with the greatest number is the winner.

Examples:

3 + 2 x 4 = ___

a. (3 + 2) x 4 = ___ | 3 + (2 x 4) = ___
5 x 4 = 20 | 3 + 8 = 11

7 – 5 + 2 = ___

b. (7 – 5) + 2 = ___ | 7 – (5 + 2) = ___
2 + 2 = 4 | 7 – 7 = 0

Version 2 of The Greatest Number:

Write four 1-digit numbers. (3, 4, 5, 6) Ask each player to insert +, –, x, or ÷ between these numbers to give a solution. The player whose answer is the greatest number is the winner. Possible solutions for 3, 4, 5, 6.

a. (3 + 4) x (5 + 6) | b. 3 + (4 x 5) + 6
7 x 11 = 77 | 3 + 20 + 6 = 29

c. (3 x 4) + (5 x 6) | d. (3 x 4) x (5 x 6)
12 + 30 = 42 | 12 x 30 = 360

Version 3: Change the title to The Least Number and let the winner be the player whose answer is the least number.

5. To introduce the process of multiplying by 10, suggest that your child solve the first six equations below on a calculator. Discuss the pattern that resulted: "How did each number change when you multiplied by ten?" (A zero was annexed or attached on the right.) Have the child solve the remaining equations without the calculator, but use it to check the answers.

10 x 3 = ___ 10 x 8 = ___ 10 x 12 = ___
10 x 25 = ___ 10 x 42 = ___ 10 x 232 = ___
10 x 40 = ___ 10 x 300 = ___ 10 x 2562 = ___

6. To introduce multiplication by multiples of 10 (20, 30, 40...) suggest that your child solve the first three equations in the following columns with the calculator, record the answers, solve the remaining equations without the calculator and check the answers with the calculator.

3 x 2 = ___	4 x 6 = ___	3 x 12 = ___
30 x 2 = ___	4 x 60 = ___	30 x 12 = ___
3 x 20 = ___	40 x 6 = ___	3 x 120 = ___
3 x 2000 = ___	400 x 6 = ___	3 x 1200 = ___
300 x 2 = ___	4 x 600 = ___	300 x 12 = ___

Ask: "How did you find the products?" (Multiplied the non-zero digits and annexed the number of zeros that are in the factors.)

7. Ask your child to help you plan for grocery shopping in the following ways:
   a. Find and write the prices of some grocery items on your list as they are listed in the grocery ads;
   b. use the calculator to find the total cost of these items; and
   c. check the reasonableness of the answer by rounding each price to the nearest dollar and estimating the sum as shown above. Note: ≈ is the symbol for is approximately equal to.

$1.79	≈	$2.00
2.43	≈	2.00
0.87	≈	1.00
1.69	≈	2.00
1.37	≈	1.00
4.25	≈	4.00
$12.40	≈	$12.00

8. Find the price of an item in the grocery ads and ask your child to use the calculator to find the price of five of these items and to estimate the reasonableness of the answer by rounding the price and multiplying.
   Example: 5 cans of apricots @ $1.08.

$1.08	≈	$1.00
x 5	≈	x 5
$5.40	≈	$5.00

9. Write the length of each side of a lot to determine the amount of fencing that you need. Ask the child to use the calculator to find the total of the lengths, round the lengths to the nearest multiple of ten, and estimate the sum to check the answer.

Summary

The many interests of children provide great opportunities to involve them in interesting problem-solving activities that will improve their math skills. Game scoring, collecting items, and solving puzzles encourage counting and computation. Children practice measuring skills by finding the size of playing areas, their own personal measures, and those of family members and friends. These same activities can motivate children to use calculators, and, in turn, guided use of the calculator can improve their math skills.

Be alert to your child's changing interests and ever-expanding math skills and adjust the problem-solving situations that you present to match these new interests and skills.

11

GAMES

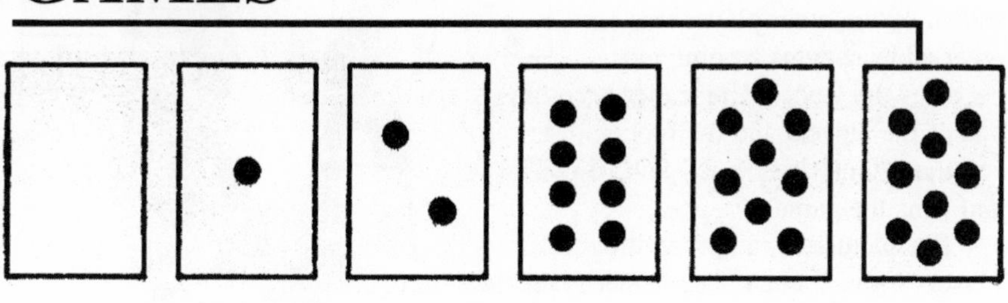

Games are so universally enjoyed by children that they can be used for many developmental purposes. With your guidance, playing games can help your child learn to interpret directions, to follow rules, and to cooperate with other people. The games in this chapter will be used for these purposes and to develop mathematical skills in a relaxed and happy atmosphere; an atmosphere that will lead the child to develop a positive attitude toward mathematics.

All of the games in the following pages are referred to in Chapters 1 through 10. Follow these references to determine when your child is ready for a specific game. While some may be won by luck, all will provide practice for the mathematical processes. Because people have different interests, your child will enjoy some games more than others. There are many games for the same purpose, so try each one and let the child choose the ones to repeat

Back to Back

GAME 1:

Purposes:

a) To practice addition and subtraction facts with sums to 18; and

b) to practice solving number sentences in which there is an addend missing.

Examples: 4+ _ =9: _ +3=7.

Materials: Two sets of numeral cards, 2 to 9. (One set for each of two players.)

This game is for a leader and two players. Play it with your child after he has studied all addition facts with sums to 18.

Procedure: Two players sit back-to-back so that neither can see the numeral shown by the other. At a signal ("Go") from the leader, each player holds up a numeral card for the leader to see. The leader names the sum (result of addition) of the two numbers shown by the players. The player that first names the number shown by the other player (the other addend) wins a point and becomes the leader. The leader becomes one of the players and the process continues until one player gets 10 points and wins the game.

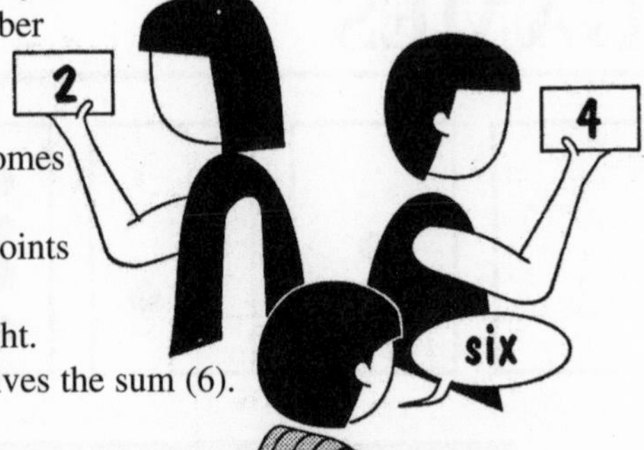

See pictured example to the right. Players show 2 and 4. The leader gives the sum (6).

BACK TO BACK, GAME 2:

Purposes:

a) To practice multiplication and division facts with products to 81;and

b) to practice solving multiplication sentences in which there is a factor missing.

Examples: 3x_=18; _x5=20.

Materials: Same as for Game 1.

Play this game after your child has studied all multiplication and division facts with products to 81.

This game is played in a manner similar to Back to Back, Game 1, with the following changes:

a) The leader states the product of the two numbers; and

b) each player tries to guess the factor shown by the other player. See pictured example at right. Players show 2 and 4. The leader gives the product (eight).

Bean Bag Toss

GAME 1:

Purpose: To practice using vocabulary: in, on, under, beside, over, behind, any other positions.

Materials: Chair, bean bag, and box.

Procedure: Place a large box on a chair a few feet from the players. Each player, in turn, tosses the bean bag at the box and describes where it lands: in the box, on the chair, under the chair. The players get one point for throwing the bean bag in the box, and one point for using one of the vocabulary words previously listed above to describe where the bean bag landed. Note: A player gets two points for tossing the bean bag into the box and correctly describing where it landed. He gets one point if he tosses it into the box or describes where it lands with a vocabulary word. The winner is the first player to get ten points or any other total that is agreed upon.

BEAN BAG Toss, GAME 2:

Purpose: To practice using vocabulary: nearer, farther, nearest, farthest (with three or more players)

Materials: Bean bag for each player, string or yardstick.

Procedure: Use string, a yardstick or the edge of a carpet for a starting line. Place another string or yardstick several feet away for the goal line. the distance depending on the ability of the players. Have two players stand at the starting line, toss the bean bag, and see who can toss it nearer the goal line. After each player has had a turn, ask each player to describe the position of his or her bean bag as nearer to or farther from the line than the other player's bag. The player whose bag is nearer the line gets a point. The winner is the first player to get ten points.

Follow the same procedure with three or more players and ask each player whose bag is nearest to or farthest from the goal line to describe the position of his bag.

Building the Greatest Number

Purpose: To read and compare numbers with nine digits.

Materials: Two sets of numeral cards. 0 to 9, or a set of playing cards with the face cards removed, and a place value chart like the one shown here for each player.

Procedure: Shuffle the cards and turn them face down. Players will take turns turning the top card face up. When a card is turned over. each player copies the number in any empty column of this chart without showing the other players.The object of the game is to form a number of higher value than any other player.When nine cards have been turned faceup, each player reads his or her number aloud and then compares the numbers to

Millions			Thousands			Units		
hundreds	tens	ones	hundreds	tens	ones	hundreds	tens	ones

see whose number is greatest. The player whose number is the greatest wins a point. Play continues until one player gets ten points and is the winner of the game.

Example: Player 1: 763217804
Player 2: 876403721
Player #2 is the winner.

Buy or Sell

Purpose: To practice making change after a purchase.

Materials: a) Ten 3-inch by 5-inch file cards with a direction for buying or selling an item written on each:

Examples:

Sell 1 crayon	Buy 1 spoon	Sell 1 Eraser and 1 Pencil	Sell 1 toy	Buy 1 top and Sell 1 eraser.	Buy 1 Pencil.

b) fifteen to twenty small items labeled with prices less than one dollar;

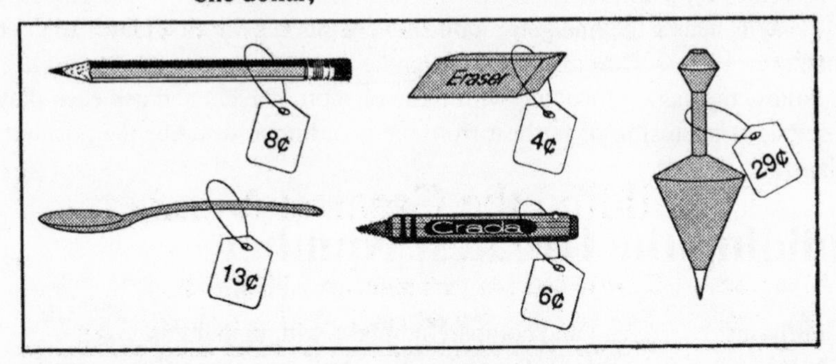

b) fifteen to twenty small items labeled with prices less than one dollar,
c) three quarters for each player, and
d) a box of change.

Procedure: Place the cards face down between the players with the labeled items beside the cards. Give each player three quarters to begin the game. Place a box of change, the bank. in the center. The players lake turns drawing a card and following the directions by buying an item from, or selling an item to, the other player. A coin may be exchanged at the bank for change by the player that is selling the item. If either player makes an error in computing the amount of the purchase or the amount of change to be received, he must pay the bank a penny as a penally. When all of the cards have been drawn and all purchases completed, the player with the most money is the winner and gets to keep his money. The loser must return his money to the bank.

Buzz

Purpose: To practice any multiplication facts.
Materials: None.
Procedure: This game can be played anywhere. It is fun to play in the car on a trip. Player #1 names a number, 1 to 9, and begins the counting by saying "one." The other players, in turn, say the next number in the counting order. except for the pre-selected number. The word "Buzz" is substituted for this number and its multiples.

Example: 5 players.

Player #1 names 4 as the Buzz number, then begins to count: "One." Player #2: "Two." Player #3: "Three." Player #4: "Buzz." Player #5: "Five-." Player #1: "Six." Player #2: "Seven." Player #3: "Buzz." etc. Any player who makes a mistake drops out of the game. Play continues until only one player, the winner, remains in the game.

To make the game more difficult and more interesting, add a second number that cannot be said. Players will say "Bang" for this number and its multiples. Let's use 6 as an example: 1, 2, 3. Buzz, 5, Bang. 7 Buzz, 9, 10. 11, Buzz Bang, 13.14.15, Buzz. 17. Bang, 19. Buzz. 21.22,23, Buzz Bang...

Checker Cover Up

GAME 1:

Purpose: To practice counting to ten.
Materials:

a) Two sets of picture cards with 0-10 dots, stars, or other small items on each. To prepare the cards, paste or draw the items on four-by six-inch file cards.

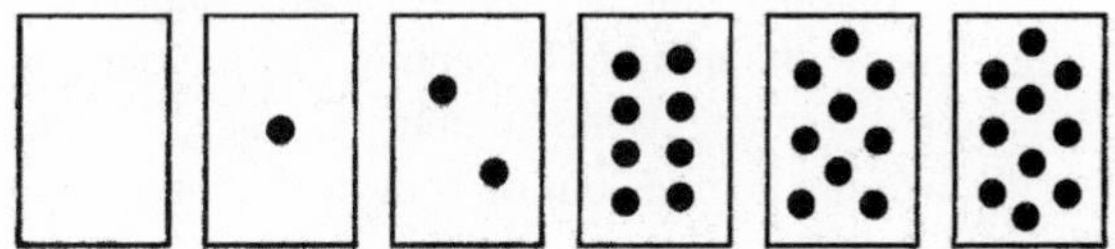

b) two sets of checkers (24 red and 24 black); and
c)a checkerboard.

Procedure: The game is for two players. Give the red checkers to one player and the black to the other. Shuffle the cards and place them, face down, beside the checkerboard. The first player draws a card. counts the dots, and places that number of checkers on the squares (red and black) on his or her side of the checkerboard. The second player follows the same procedure. The players continue, in turn, until one player has covered all of the squares in the first three rows of his side of the board. That player is the winner.

GAME 2:

Purpose: To practice recognizing numerals and counting to ten.

Materials: a) Two sets of numeral cards (0 to 10). To prepare the cards. write each numeral on a separate four- by six- inch file card.
b) Two sets of checkers (24 red and 24 black).
c) A checkerboard.

Procedure: Same as for Game 1 with each player, in turn, identifying the numeral on the drawn card and placing that number of checkers on the board.

Clean the Board

Purpose and Materials: Same as for Checker Cover Up.

Procedure: Each player places a set of 24 checkers (red or black) on one side of the board. Players, in turn, draw a card and remove as many checkers as indicated by the pictures or numeral on the card. The first player to clean his or her side is the winner

Empty My Pockets

Purpose: To practice removing objects from a set and writing the matching subtraction sentence.

Materials: Ten pennies and a pocket

Procedure: Play this game when your child begins to study subtraction sentences. Ask your child to:

a) Count as you put some pennies (3) in your pocket and write the number,
b) take some pennies (2) out of your pocket and finish writing the number sentence that tells what happened (3 - 2 = _)
c) "guess" the number of pennies left in your pocket and write the difference in the number sentence (1); and
d) take the pennies out of your pocket and count to check the difference. If the "guess" is correct, the child keeps the pennies. If not, you get the pennies.

Facts Football

Purpose: To practice addition, subtraction, multiplication, and division facts.

Materials: a) Playing field as described in Scale Drawings, Chapter 10;
b) cards with facts that your child is studying; and
c) paper football or button to represent the ball.

Procedure: Separate cards into groups according to level of difficulty and turn them face down. Label the stacks according to the amount of yardage gained by giving the correct answer: Level 1 (easiest facts)-1 yard gain; Level 2-3 yard gain; Level 3 (most difficult)-5 yard gain; Level 4-This stack will contain cards with a combination of operations on each and be used for a punt or field goal on fourth down. Examples: (3 + 4) x 5; (3 x 4) + 5; 3 + (4 x 5).

This is a game for two players (teams) and a referee.

Pitch a coin to determine which team (player) gets the ball first. That team begins on their own 20-yard line and for each play indicates to the referee from which stack to draw the fact that the team must answer to move the ball. The team has four downs in which to gain 10 yards and make a first down. On fourth down, if the team with the ball chooses a card from the level 4 stack, both teams can try to answer. If the team with the ball answers first, they have made a field goal and get 3 points. If the other team answers first, it was a punt and that team gets the ball halfway between the location of the ball and their goal line. If a team loses the ball by failing to make four downs, the opposing team takes the ball where it is located. After a touchdown or field goal, the other team starts with the ball on its own 20-yard line. The length of the game can be determined by a time limit (5 minutes per quarter) or by a number of points... Play until one team gets 18 points.

Facts Relay Race

GAME 1:

Purpose: To practice addition facts with sums to ten.

Materials: Cards with addition facts, sums to ten.
Numeral cards, 0 to 10 (optional)

Procedure: This is a game for two players and a leader. The leader shows an addition flashcard and the player that gives the sum first, orally or with a numeral card. wins a point The player with the most points at the end of play is the winner.

FACTS RELAY RACE, GAME 2:

Purpose: To practice addition facts with sums to 18.

Materials: Addition fact cards, sums to 18. | 7-"-8 |

Procedure: Same as for Game 1.

FACTS RELAY RACE, GAME 3:

Purpose: To practice subtraction facts, sums to 10 or 18.

Materials: Subtraction fact cards, sums (top numbers) ten or less or between 10 and 18.

Procedure: Same as for Game 1. Player gives the difference.

FACTS RELAY RACE, GAME 4:

Purpose: To practice multiplication facts.
Materials: Multiplication fact cards.
Procedure: Same as for Game 1. Player gives the product

FACTS RELAY RACE, GAME 5:

Purpose: To practice division facts.
Materials: Division fact cards.
Procedure: Same as for Game 1. Player gives the quotient

Fewest Coins

Purposes: a) To practice counting by a combination of ones, fives, tens, or twenty-fives; and
b) to practice making change.

Materials: A set of coins for each player, pennies, nickels, dimes, quarters, and half dollars.

Procedure: This is a game for a leader and two or more players. The leader names an amount of money. The player that displays that amount with the fewest coins is the winner. If both, or all, players use the same number of coins, the player that makes the display first is the winner.

Examples:

	Amount named by leader	Fewest coin display:
a.	37¢	1 quarter, 1 dime, 2 pennies (4 coins)
b.	76¢	1 half dollar, 1 quarter, and 1 penny (3 coins)

Fill My Pockets

GAME 1:

Purpose: To practice joining sets of objects and writing an addition sent ence with the sum. Ten pennies and a pocket Use this game
Materials :when your child begins to study addition. Ask your child to:
Procedure:

a) Count as you put some pennies (2) in your pocket and write t he numeral (2);

b) count as you put some more pennies (1) in your pocket and finish writing the number sentence that tells what happened (2 + 1 = _);

c) "guess" the number of pennies in your pocket and write the sum in the number sentence (2 + 1 = 3); and

d) take the pennies out of your pocket and count to check. If the guess is correct, the child keeps the pennies. If not, you keep the pennies.

FILL MY POCKETS, GAME 2:

Purpose: To practice joining equivalent sets of objects and writing a matching multiplication sentence.

Materials: Several pennies and a pocket Procedure: Ask your child to:

a) Put four sets of three pennies in your pocket and write a multiplication sentence to tell what happened (4x3= _);

b) "guess" the number of pennies in your pocket and write the product in the number sentence (4x3= 12); and

c) take the pennies out of your pocket and count to check the product.

If the "guess" is correct, the child keeps the pennies. If not, you keep the pennies. Continue with other equivalent sets of pennies. Have the child record each sentence with the product.

Guess the Number

Purpose: a) To practice counting up to ten objects; and
b) to practice addition and subtraction.

Materials: Any object similar to the ones named below.

Procedure: As you go about your regular duties and activities, you can supply the objects named here and others to be counted and ask questions to encourage the child to discover sums and differences.

a) Say: "Put one red apple on the tray. Now, if you put one yellow apple on the tray, how many will there be?" After the child has "guessed" the number, have him put the yellow apple on the tray and count to check the answer. Say: "One and one makes two." Place two. three, or four apples on the table and ask the child to give you a "story" with a ques tion for you to guess the answer. Each time, a player that gets the right answer gives the next story.

b) Place 3 apples on the tray. Say: "If we take two apples from the tray, how many will be left?" After the child guesses, remove two apples and ask him to count to check the guess. Encourage the child to give the next story for you to guess the answer.

c) Watch out for this one! For a little enrichment ask the old question: "Take 2 apples from 3 apples and what do you have?" (2 apples. You took 2 so you have 2 and 1 is left on the tray.)

Hide the Thimble

Purpose: To identify left, right, higher, lower, forward, backward.
Materials: A thimble or other small object.
Procedure: Hide the thimble. Place it so that it can be seen from some positions without moving anything. Have your child look for it as you give clues. "Move to your left. Look a little higher. Move to your right. Lower. Much lower. Move forward." Add incentive by hiding a small treat, a penny or a nickel, instead of a thimble.

High or Low

Purpose: To practice addition, subtraction, multiplication, or division.
Materials: Deck of playing cards with face cards removed.
Procedure: This is a game for 2 or more players. On a sheet of paper, draw boxes to show the kind of exercise you want to practice. Before starting, decide whether you are playing for highest or lowest answer. Shuffle the cards and place them face down. One player turns the first card face up and each player decides where to write that numeral in the boxes. Once it is written, it may not be moved. The other player turns over the second card and, again, each player writes that numeral in one of the boxes. After four cards have been turned over and recorded in the boxes, each player finds the answer to his problem. The player with the highest (or lowest, depending on the game you chose) answer, scores a point. Example: The following cards were drawn: 8, 3, 6, 2.

Ex. □ □ □
× □

Player # 1.	Player # 2.	Player # 3.
Ex. [8] [3] [6]	Ex. [6] [3] [2]	Ex. [8] [2] [3]
× [2]	× [8]	× [6]
1 6 7 2	5 0 5 6	4 9 3 8

Player #2 won this game. The strategy will change, depending on whether you are playing for high or low answer.

Play the above game with any of the following operations:

□ □ + □ □ □)□ □ □ □ □ − □ □

I Spy

GAME 1:

Purpose: To practice using positional words: near, far, beside, above, below, on.

Procedure: Choose any object in the room that is in plain sight. Give clues that describe where the object is. Examples: You have selected the toaster. Say: "I spy something that is near the refrigerator. (Let the child guess. If the guess is incorrect, continue with clues.) It is/or from the door. (Child guesses.) It is beside the can opener. (Child guesses.) It is on the cabinet." Give a reward when the child guesses the object. Let the child be the "spyer" for the next game and provide the clues for you.

I SPY, GAME 2:

Purpose: To practice using comparison words for measurement: larger, smaller, heavier, lighter: longer, shorter, taller.

Procedure: Play this game by the same rules as Game 1 but use comparison words in the clues.

Example: You have chosen a chair. Say: "It is smaller than the sofa. It is larger than this table. It is heavier than the book."

Least

Purpose: To identify the number.0to9.that is least

Materials: 3 sets of numeral cards, 0 to 9.

Procedure: This is a game for three or more players. Shuffle the cards and place them face down. Each player draws a card and turns it face up. The player that draws the number that is least takes all of the drawn cards. Play continues until all cards have been drawn. The player with the least number of cards wins.

Less

Purpose: To identify the number, 0 to 9, that is less. Materials: Numeral cards 0 to 9.

Procedure: This game is for two players and is played similarly to Least with the following adjustments:

a) The player with the numeral that names less than the other numeral keeps the two cards.

b) The player with less cards at the end of play wins.

Match Up

GAME 1:

Purpose: To practice matching the numerals, 1 to 10, with sets of the same number.

Materials: One set of numeral cards, 0 to 10, and one set of picture cards as shown and described in materials for Checker Cover Up.

Procedure: Select eight numeral cards and the matching set cards (16 cards). Turn the cards face down, in mixed order in four rows, the numerals in two rows and the sets in the other two rows as shown on the right. The first player turns one numeral card and one set card face up. If the numeral matches the number in the set of shapes, the player keeps the two cards. If the two cards do not match, the player turns them face down in the original positions. The other player follows the same procedure. Play continues, in turn, until the players have removed all of the cards. The player with the most cards at the end of the game is the winner.

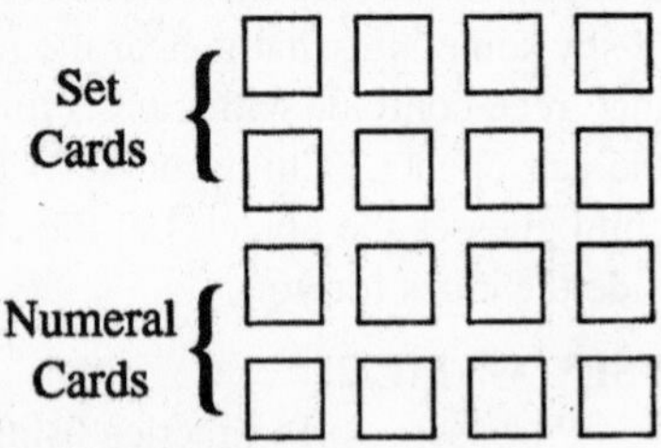

Set Cards

Numeral Cards

MATCH UP, GAME 2:

Purpose: To practice addition facts.

Materials: Eight addition fact cards and eight cards with the matching sums. Ex. [2 + 4] [6]

Procedure: Same as for Game 1. Match each addition fact with its sum.

MATCH UP, GAME 3:

Purpose: To practice subtraction facts.

Materials: Eight subtraction fact cards and eight matching difference cards. Ex. [7 – 2] [5]

Procedure: Same as for Game 1. Match each subtraction fact with the difference.

MATCH UP, GAME 4:

Purpose: To practice multiplication facts.

Materials: Eight multiplication fact cards and eight matching product cards. Ex. [5 x 7] [35]

Procedure: Same as for Game 1. Match each multiplication fact with a product.

MATCH UP, GAME 5:

Purpose: To practice division facts.

Materials: Eight division fact cards and the matching quotient cards.

Procedure: Same as for Game 1. Match each division fact to a quotient.

Ex. [18 ÷ 3] [6]

More

Purpose: To identify the number. 0 to 9, that names more. Materials: Numeral cards 0 to 9.

Procedure: This is a game for two players and is played similarly to Least with the following changes:

a) The player with the numeral that names more than the other numeral keeps the two cards.

b) The player with more cards at the end of play wins.

Most

Purpose: To identify the number, 0 to 9. that names the most. Materials: Numeral cards, 0 to 9.

Procedure: This game is for three or more players and is played similarly to the game of Least with these changes:

a) The player that draws the number that names the most keeps the cards.

b) The player with the most cards at the end of play wins. When discussing t he games Least, Less, More, or Most with children, feel free to use any of the following expressions: less than, fewer than, more than, greater than, greatest.

Mystery Box

GAME 1:

Purpose: To count to ten.

Materials: A box with lid, 10 beans or buttons.

Procedure: Put one bean in the box and ask your child how many beans are in the box. Close the lid. Say: "I will put one more bean in the box. (Put it in without letting the child see inside.) Now, how many beans do you think are in the box? Open the box and count the beans." As the child learns to count to higher numbers, increase the number of beans put in the box and always add one more before the "guess." Give the child a point for every correct guess and give yourself a point for every incorrect guess the child makes. The one with more points at the end of play is the winner

.

MYSTERY Box, GAME 2:

Purpose: To practice addition facts with sums to 10.

Materials: Same as for Game 1.

Procedure: Place a number of beans (3) in the box as the child counts them. Show another set of beans (2) and say: "How many beans will be in the box when I put in these? Let's count. Three are in the box so one more will be four (drop 1 into the box) and one more will make five. (Drop another into the box.) Let's count and see if 3 and 2 more make 5." Encourage the child to count as you put the additional beans in the box, beginning with the number in the box and saying the next number for each bean you add. Score as in Game 1.

MYSTERY BOX, GAME 3:

Purpose: To practice subtraction facts with sums to 10.

Materials: Same as for Game 1.

Procedure: Place a number of beans (3) in a box as the child counts them. Say: "How many beans do you think will be in the box if I lake out 2? Let's try it. I'll take two out and how many are left? (1) Let's check. Yes, we had three, took away two, and had one left." Continue with other numbers of beans (ten or less) and ask the child to count backward as you take beans from the box. Score as in Game 1.

MYSTERY BOX, GAME 4:

Purpose: To practice writing and solving addition facts with sums to 10.

Materials: Same as for Game 1 and paper and pencil.

Procedure: Show a number of beans (5), have the child count and record the number as you place them in the box. Show another set of beans (3). Place these beans in the box and have the child complete a number sentence that tells what happened. (5 + 3 = _) Urge him to guess the number of beans in the box and write the number on the blank. Open the box and check the answer. Reverse roles. Let the child show the beans and you write the sentence and answer. Give a point to each player that writes the correct sentence. The winner is the one with the most points at the end of play.

MYSTERY BOX, GAME 5:

Purpose: To practice writing and solving subtraction facts with sums to 10.

Materials: Same as for Game 4.

Procedure: As the child watches, place from one to ten beans (8) in the box and ask the child to write the number. Without letting him see inside the box, remove some of the beans (3) and place them beside the box. Have the child write the number sentence to show what happened (8 - 3 = _) and guess and write the number that is still in the box. Open the box and have the child count to check. Reverse roles. Score as in Game 4.

MYSTERY BOX, GAME 6:

Purpose: To practice multiplication facts.

Materials: Box with lid, paper and pencil, enough small items to make sets to fit the multiplication facts.

Procedure: Show the number of beans that you will put in the box each time and have the child record the number (2), count the number of times you put this many beans in the box (3), and write the appropriate multiplication number sentence. (3x2= _) Ask the child to "guess" the number in the box, write the guess on the _, and then open the box to count the number of beans. If the guess is correct, that player gets a point. Reverse roles. Determine the winner by counting points at the end of play.

Name the Number

Purpose: To practice naming the number of items grouped in tens and ones to 99. Materials: 99 sticks (tongue depressors or popsicle sticks); 9 rubber bands or pieces of string; 2 sets of numeral cards 0 to 9, one set for each player.

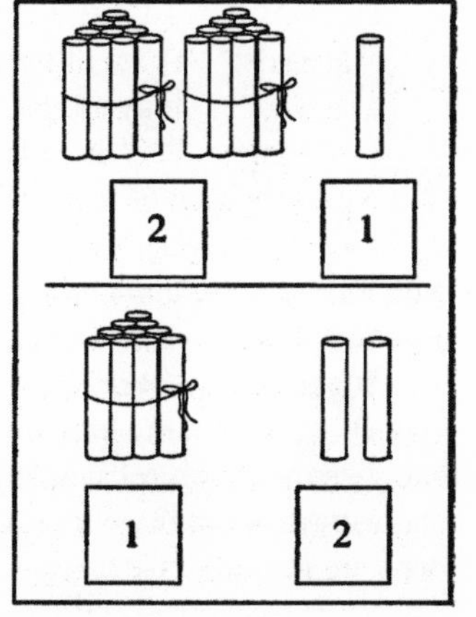

Procedure: This is a game for two players and a leader. The leader shows some bundles of ten sticks (2) and some individual sticks (1). Each player uses his or her numeral cards to form the number that tells how many sticks. The first player to form the number wins a point. Have ten numbers formed and the player with the most points wins. Stress that the number on the left tells the number of tens and the number on the right tells the number of ones.

Number Relay

GAME 1:

Purpose: To practice forming numbers above 99.

Materials: Two sets of numeral cards, (0 to 9).

Procedure: This game is for a leader and two players. Give each player a set of the numeral cards. Name a number orally. The first player to form the number wins a point At the end of play the player with the most points is the winner.

Example:

Leader says: "Five thousand four hundred six."

Players four the number 5406

The leader should not give a number in which a digit is repeated because each player only has one of each numeral.

NUMBER RELAY, GAME 2:

Purpose: To write the numbers above 99.

Materials: Writing material for each player.

Procedure: The leader names a number above 99. The first player to write the number wins a point. At the end of play, the person with the most points is the winner. Important! Show thai a comma comes between hundreds and thousands. between hundred thousands and millions, etc. A comma follows each third number to the left!

Numeral Race

Purpose: To identify numerals 0 to 10.

Materials: A set of numeral cards. 0 to 10, for each of the two or more players.

Procedure: This is a game for two or more players and a leader. Give each of the players a set of numeral cards, 0 to 10. Each player spreads the cards face up in his or her playing area. The leader asks a question that is to be answered with one of these numerals. The first player to hold up the correct numeral wins a prize-a penny, a mint, a peanut.

Suggested questions: How many fingers do you have? How many cars do you have? How many legs are on the table? How many tails are on our dog? How many sides are on a triangle? How many toes do you have on one fool? How many crayons are in your crayon box? (Be sure there are ten or less.) How many lions are in our house?

Pay the Piper

GAME 1:

Purpose: To practice addition facts with sums to ten or eighteen.

Materials: Addition flashcards (commercial or home made) with sums 0 to 10 or 11 to 18. One penny for each card; checkers or snap beads of two colors.

3 + 2 | 6 + 4 | 4 + 3 | 5 + 4

Procedure: Put the pennies in the center of the playing area. Show a flashcard. The player takes one penny from the pile when he gives the sum before you count silently to five. When the player is unable to give the sum in time, you take a penny from the pile. After all of the cards have been shown, let the child use two colors of snap beads or checkers to build each sum that he missed.

Example: 3 + 2

Model:

Have the child look at the model, read the fact and give the sum. (Three plus two equals five.) For practice, mix the flashcards that the child needs to study and let the child place each card beside the model of that fact.

Put pennies that you took from the previous game of Pay the Piper back into the center of the playing area. The number of pennies should match the number of facts that the child needs to learn. Play the game as before but use only those sums that the child missed. Other treats may replace pennies-pieces of fruit, cookies, etc.

PAY THE PIPER, GAME 2:

Purpose: To practice subtraction facts, sums to 10 or 18. Materials: Subtraction flashcards; one penny for each flashcard; snap beads or checkers to model each fact.

Procedure: Adjust Game 1 by using subtraction flashcards. Have the child model the facts that he or she misses by building the model of the top number and then removing the number to be subtracted to find how many are left.

PAY THE PIPER, GAME 3:

Purpose: To practice multiplication facts.

Materials: Multiplication flashcards; small items to model facts; one penny for each flashcard.

Procedure: Play the game as described for Game 1, but use multiplication flashcards. After the cards have been shown, urge the child to model any fact that he missed.

Example: 5 x 3 = _

Model:

●●●●●
●●●●●
●●●●●

PAY THE PIPER, GAME 4:

Purpose: To practice division facts.

Materials: Division flashcards; one penny for each flashcard; 81 small objects for modeling the facts.

Procedure: Play the game like Game 1 but use division flashcards. Urge the child to model any facts that he misses.

Example: 15 + 3 = _. Place 15 small objects together and sec how many sets of three can be made.

Secret Sum

Purpose: To practice addition facts.

Materials: None.

Procedure: This game can be played while you are involved in other activities. The leader decides on a secret sum greater than nine. The player names an addend from one to nine and the leader tells the other addend that would make the secret sum. The player tries to guess the sum. If the guess is incorrect, the player continues to name addends and each time the leader tells the matching addend that will give the secret sum.

Example: The player says: The leader says:

The player says:	The leader says:
Six.	Five.
Four.	Seven.
Nine.	Two.

Continue in this way until the player guesses the secret sum of eleven. The player becomes the leader and decides on a secret sum.

Secret Rule

Purpose: To practice addition, subtraction, multiplication, and division facts.

Materials: None.

Procedure: This game can be played while washing dishes, taking a ride. or performing other activities. The leader decides on a number to add to, subtract from, multiply by, or divide by any number that the player names. The leader keeps the number and the process a secret. The player names a number from one to nine. Then the leader tells the number that would result after adding, subtracting, multiplying, or dividing his number, depending on the rule. If the player can guess the leader's secret process and number, he wins a point. If not. the player names another number from one to nine. Again, the leader gives the number that would result from applying the same secret rule. The player continues as before until he is able to name the secret rule. The player can then become the leader and think of the secret rule.

Example # 1.

Player says:	Leader says:
Four.	Eight.
Five.	Ten.
Eight	Sixteen.

Secret rule: Multiply by two.

Example #2.

Player says:	Leader says:
Six.	Ten.
Three.	Seven.
Nine.	Thirteen.

Secret rule: Add four.

Example #3. (Tell the player to give numbers between 5 through 10, and keep the number subtracted five or less. This will keep the answer from being less than zero-a negative number.)

Player says:	Leader says:
Seven.	Six.
Nine.	Four.
Three.	Six.

Secret rule: Subtract three.

Example #4. Player says:	Leader says:
Nine.	Three.
Seven.	Two and one third or T wo, remainder one.
Six.	Two.

Secret rule: Divide by three.

Shapes

Purpose: To identify circles, triangles, rectangles, and squares by feel.

Materials: Cut circles, triangles, squares and rectangles from cardboard or heavy paper; paper bag.

Procedure: Place all of the shapes into a paper bag. The first player reaches into the bag, names a shape, and pulls out that shape. No peeking! That player gets a treat-a penny, a mint, a big hug from Mommy-if he pulls out the named shape. The second player does the same.

Variation: One player may name the shape for the other player to pull from the bag.

Show the Number

Purpose: To practice using tens and ones to model a number from 10 to 99.

Materials: Two sets of numeral cards (0 to 9); nine bundles of ten sticks and nine individual sticks for each player. (Toothpicks, tongue depressors, ice cream sticks.)

Procedure: This is a game for two or more players and a leader. With the numeral cards the leader shows a numeral from 10 to 99. The first player to model that number with sticks, wins a point

After ten numbers, the player with the most points wins.

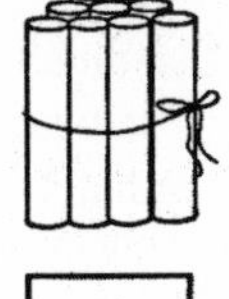

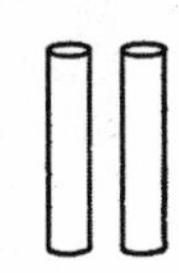

2

Simon Says

GAME 1:

Purpose: To identify positions: above, over, under, beside, forward, backward, on, behind, up, down, left, right

Materials: None.

Procedure: If it's been a long while since you've played Simon Says, the rules are simply that players follow only those directions that are preceded by "Simon Says." If Simon doesn't say it, don't do it! In a group situation, players are eliminated as they are caught doing something that Simon didn't tell them to do and the last person left in the game becomes the leader for the next game. If you are playing with only one child, just laugh together about "catching" him and make a supportive comment such as: "It took me a long time to catch you that time!" Examples of directions for developing the use of the words listed above:

Simon says put your hands above (or over) your head.
Simon says put your hands behind your back.
Simon says take one step forward.
Take a step backward. (Whoops! Simon didn't say it)
Simon says put your finger beside your nose.
Put your hands on your knees.
Simon says put your thumb under your chin.
Simon says thumbs up.
Thumbs down.
Simon says put your right hand in the ail:
Simon says stand on your left fool.
Put your left hand on your right knee.
Take one step to your right.
Simon Says take a step to the left.

Have fun by using a series of directions to:

a. Maneuver the player right out an open door;
b. get the player to bring something to you; or
c. assist the player in finding a treat.

SIMON SAYS, GAME 2:

Purpose: To count one to ten objects or one to ten movements.

Materials: None.

Procedure: Use directions such as the following:

Simon says stand on one foot
Hold up one hand.
Simon says take two steps forward.
Take three hops backward.
Clap your hands/5 times.
Simon says take five elephant steps.

SIMON SAYS, GAME 3:

Purpose: To recognize the numerals 1 to 10.

Materials: Numeral cards 1 to 10.

Procedure: Play the game like Game 1 using directions that involve the numbers one to ten. Instead of saying the number, show the numeral to indicate the number.

Example: Simon says take (show numeral card 3) kangaroo leaps.

Store

GAME 1:

Purpose: a) To identify prices marked with the cent sign ((); and b) to count from one to nineteen. Materials: Any items around the house labeled with prices of 1¢ to 19¢; pennies.

Procedure: Explain ¢ as meaning cents. Let the child use pennies to buy items that are labeled with prices of 1¢ to 19¢. (An apple, an orange, a balloon, a box of crayons, a pencil, etc.) Explain that he will be given a chance to earn the money back. Use the money later as a prize for winning a math game.

STORE, GAME 2:

Purposes: a) To count ten to ninety-nine;
b) to model numbers ten to ninety-nine with tens and ones; and
c) to use money that includes quarters and half-dollars.

Materials: Items around the house labeled with prices from lO¢ to 99¢; pennies and dimes.

Procedure: Let the child use pennies and dimes to buy items, one at a time, that are labeled. Include quarters and half-dollars after the child has learned the value of these coins (Chapter 4)

STORE, GAME 3:

Purpose: To find the sum of two numbers, one to ninety-nine. Materials: Items labeled as in Game 2; pennies, dimes, and one dollar bills.

Procedure: Let the child use pennies, dimes, and dollars to buy two or more items by finding the sum of their prices and giving the correct coins and bills.

STORE, GAME 4:

Purposes: a) To subtract 2- and 3-digit numbers from a 3-digit number. and
b) to find the amount of change due from a coin or a bill when paying for a purchase.

Materials: Same as for Game 3.

Procedure: Encourage the child to buy the items as in Game 3. pay with a coin or bill of greater value than the purchase, and subtract to check the amount of change he received after the purchase.

Example:

$$\begin{array}{r} \$5.00 \\ \underline{-3.98} \\ 1.02 \end{array}$$

Tic Tac Toe

Purpose: To practice addition, subtraction, multiplication, and division facts.

Materials: Writing materials.

Procedure: Draw a Tic Tac Toe frame on paper, a magic slate, or chalkboard. Write an addition fact (subtraction, multiplication, or division) in each section. The first player gives a sum (difference, product, or quotient) of one of the facts orally. If it is correct, that player writes (he sum in the proper place and puts 0 in the section.

The second player chooses a fact. gives the sum orally, and if it is correct, writes an X in that section. The first player to have three of his marks. X or 0. in a row vertically, diagonally, or horizontally, is the winner.

2 + 3 = 5 (O)	4 + 5	3 + 3 = 6 (X)
4 + 3 = 7 (X)	2 + 5 = 7 (X)	2 + 2
5 + 3 = 8 (O)	6 + 3 = 9 (O)	3 + 1 = 4 (O)

The Trail Game

Purpose: To practice addition, subtraction, multiplication, and division facts.

Materials: The playing board from the next page; a button of a different color for each player, addition, subtraction, multiplication, or division flashcards as used in Pay the Piper.

Procedure: To practice one operation (addition, subtraction, multiplication, or division), use flashcards showing only that operation. Combine two or more operations to train the child to check the sign before giving the answer. The first player places a button on start, draws a flashcard from the face-down stack, tells the answer, and moves the button as many spaces (following the arrows on the game board) as named by the answer to the fact. The second player follows the same procedure. Play continues until one player reaches Home with his button and is the winner. If one button is in the path of the one being moved, the player jumps the button and counts it as one move.

The Trail Game Playing Board

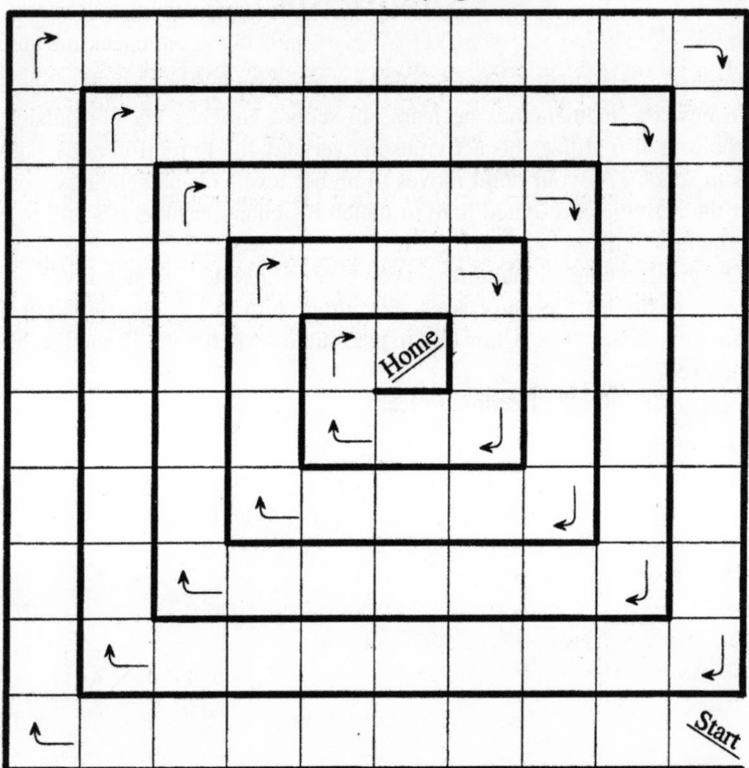

Conclusion

The emphasis on realistic problem solving throughout this book helps your child realize how the mathematics he learns in school enriches his life outside the classroom. It also utilizes his activities in everyday life to reinforce the skills he learns in school. As your child moves to higher levels of mathematics, you can adjust the activities presented here to match his changing interests and to reinforce the new skills he will encounter.

The greatest educational gift you can give your child is to stay involved and interested in his learning processes. Encourage him to ask questions but don't give answers too quickly. Allow him to pleasure of experimenting and finding his own answers.

Give your child the freedom to learn.

Help Your Child Excel in Math

Notes:

Frederick Fell Publishing

Notes:

Notes: